LANGUAGE IN THE MATHEMATICS CLASSROOM

TALKING, REPRESENTING, RECORDING

RACHEL GRIFFITHS and **MARGARET CLYNE**

Heinemann
Portsmouth, NH

For Marcus and Ruth.
RG

For Margaret and Michael, Regena and Daniel.
MC

ACKNOWLEDGMENTS

We would like to thank the teachers, parents, and children at the following schools, who worked with us on the activities described in this book: Antonio Park, Ashburton, Bayswater West, Camberwell, Donburn, Eastmoor, Ferntree Gully, Ferny Creek, Kerrimuir, Kangaroo Flat, Knoxfield, Lysterfield, Marlborough, Mooroolbark, Regency Park, Syndal, The Basin, Upper Ferntree Gully, Wattle View, Waverley Park Primary Schools.

First published in 1994
Heinemann
A division of Reed Elsevier Inc.
361 Hanover Street
Portsmouth, NH 03801-3912

Offices and agents throughout the world

Library of Congress Cataloguing-in-Publication Data

Griffiths, Rachel
Language in the Mathematics classroom: talking, representing, recording / Rachel Griffiths and Margaret Clyne.
p.   cm.
Includes bibliographical references and index.
ISBN 0-435-08366-X (pbk)
1. Mathematics – study and teaching. 2. Language and education.
I. Clyne, Margaret. II. Title.
QA11.G823   1994   94-35141
510'.7–dc20   CIP

Produced simultaneously in the United States
by Heinemann and in Australia by
Eleanor Curtain Publishing
906 Malvern Road
Armadale, Australia 3143

Production by Sylvana Scannapiego, Island Graphics Pty Ltd
Cover and text design by David Constable
Edited by Ruth Siems
Cover photographs by Sara Curtain
Page make-up by Patricia Tsiatsias
Printed in Australia by Impact Printing Pty Ltd

CONTENTS

ACTIVITY OVERVIEW

Note that all activities include spoken language: discussion. This is therefore only included when the discussion is the main language mode for the activity.

ACTIVITY	FOR WHOM?	FOCUS	LANGUAGE MODES	MATHS SKILLS & CONCEPTS
CHAPTER 1 LANGUAGE MODES				
Find out what your children know p. 2 BLM 1.1	All ages	Children's prior knowledge	**Spoken language:** talking Written language: listing	General
Generating discussion on mathematics and language p. 3	Teachers and parents	Clarifying mathematics and language issues	**Spoken language:** talking Written language: listing, writing statements	General
Posing good questions p. 8	All ages	Adapting curriculum to create more open-ended problems	**Spoken language:** questioning	General
CHAPTER 2 DEVELOPING THE LANGUAGE OF MATHEMATICS				
Number messages p. 11	5–8 year olds	Ways of representing mathematical ideas	**Written language:** writing and reading mathematical instructions	Addition and subtraction
Representing division p. 13	5–10 year olds	Ways of representing mathematical ideas	**Written language/graphic communication**	Division
Division stories p. 14	10 years–adult	Finding appropriate contexts for using mathematics	**Written language:** reading mathematical text, writing stories	Division
Decimal stories p. 14	9–13 year olds	Application of concepts	**Spoken language:** explanation Written language: writing 'story'	Decimals
The wolf game p. 17 BLM 2.1	5–8 year olds	Methods of recording	**Written language:** writing, recording mathematics	Number: counting, subtraction
Homes for animals p. 18	6–9 year olds	Representing mathematical ideas	**Written language:** reading and writing narrative, recording mathematics	Subtraction

Calculator patterns p. 20	5–9 year olds	Representing and recording with different materials	**Written language:** writing numbers, explanations **Graphic communication:** drawing **Active communication:** representing with materials	Patterns in number and space
Make a book of mathematical riddles p. 22	7–13 year olds	Vocabulary	**Spoken/written language:** explaining mathematical vocabulary, writing riddles	General
Book of number words p. 23 BLM 2.2	8–13 year olds	Vocabulary/word origins	**Written language:** reading, researching mathematical words, writing explanation	Number: names of number words Space: names and properties of shapes
Topic glossary p. 23	All ages	Vocabulary	**Written language:** writing definitions	General
The four crocodiles p. 24	6–10 year olds	Vocabulary: comparatives	Written language: reading/labelling **Graphic communication:** drawing	Length, ordering
What order? p. 24 BLM 2.3	6–10 year olds	Comparing, ordering Vocabulary	**Spoken language:** discussion Written language: reading Graphic communication: interpreting drawings	Measurement, ordering
Venn diagram p. 25 BLM 2.4	6–11 year olds	Vocabulary (and, or)	**Graphic communication:** interpreting and making diagrams Written language: reading and writing Active communication: physical involvement	Logical thinking classification, organising data
Numbers in other languages p. 27 BLM 2.5, 2.6	6–13 year olds	Mathematics in other cultures	**Written language:** reading and writing numbers	Number: place value and other properties
Paper folds p. 29	6–13 year olds	Visualisation	**Graphic communication:** drawing	Space: shape and symmetry
See a number p. 30	6–10 year olds	Visualisation	**Graphic communication:** drawing Spoken language: describing	Number patterns and concepts

Estimating distance p. 30 BLM 2.7	8–13 year olds	Visualisation	**Active communication** measurement	Measurement: estimating and measuring distance
Shape and size p. 30	All ages	Visualisation	**Graphic communication:** drawing	Shape and size
CHAPTER 3 SPOKEN LANGUAGE AND COMMUNICATION				
Finding out p. 37	All ages	Focusing discussion: finding out what children know	**Spoken language:** discussion Written language: listing	General
Subtraction methods p. 40	8–13 year olds	Discussion strategy: comparing methods	**Spoken language:** explanation	Subtraction
The king who was tired of war p. 42 BLM 3.1	10 year olds–adults	Exploration and discussion in small groups	**Spoken language:** discussion Written language: recording and writing	Number: large numbers, doubling, patterns Measurement: volume, mass
Sandwich sales p. 44 BLM 3.2	6–13 year olds	Imaginative interpretation	**Written language:** writing a 'story'	Interpreting a graph
Make it like mine p. 45	5–13 year olds	Developing descriptive language skills and active listening	**Spoken language:** describing and listening	Space: shape, location Number
Family games p. 46 BLM 3.3, 3.4, 3.5	5–13 year olds and parents	Involving families	**Spoken language:** discussion Active communication: playing games	Number
Number cards p. 47	5–9 year olds	Physical involvement	**Active communication:** physical involvement	Counting ordering, place value and other number properties
Maths walk p. 48	7–10 year olds	Physical involvement, co-operative group work	**Active communication:** physical involvement	Addition
Making polygons p. 49	6–13 year olds	Physical involvement, visualisation	**Active communication:** physical involvement Spoken language: describing	Space: shape, polygons
Footsteps over the city p. 49	5–10 year olds	Physical involvement, visualisation	**Active communication:** physical involvement Graphic communication: reading maps	Space: location, maps, scale Measurement: distance
Drawing a circle p. 50	5–10 year olds	Physical involvement	**Active communication:** physical involvement Spoken language: describing, explaining	Space: properties of circles

Circumference of a circle p. 50	8–13 year olds	Physical involvement	Graphic communication: geometric drawing **Active communication:** physical involvement Graphic communication: geometric drawing	Measurement: length Ratio
A mathematical mime p. 51	10 year olds–adults	Physical involvement: mime	**Active communication:** physical involvement Spoken language: explanation	General
Maths trail p. 51	All ages	Physical involvement	**Active communication:** physical involvement Written language: reading and writing	General
CHAPTER 4 EXTENDING THE MODES OF WRITTEN LANGUAGE				
Linking mathematics and reading p. 55 BLM 4.1, 4.3	Teachers or parents	Role of mathematics in understanding what we read	**Written language:** reading, writing lists	General
Newspaper maths p. 55	9 year olds–adults	Maths in understanding what we read	**Written language:** reading, writing	General
The crow and the pitcher p. 57	7–13 year olds	Maths in stories	**Written language:** reading narrative, writing	Volume, displacement, estimation
Magic squares p. 58 BLM 4.2	8–13 year olds	Maths in historical and cultural context	**Written language:** reading	Number: number patterns, addition
Vets in Australia p. 60	8–13 year olds	Problem solving using reference texts	**Written language:** reading	Number: computation
Rule-a-code p. 61	7–13 year olds	Instructional texts using mathematics	**Written language:** reading instructions	Measurement (centimetres)
The rabbit problem p. 63 BLM 4.3	9 year olds–adults	Maths from a magazine	**Written language:** reading magazine article, writing, recording mathematics Graphic communication: diagrams, tables	Number: time
Junk mail catalogues p. 64	6–13 year olds	Maths from catalogues	**Written language:** reading catalogues, writing	Number, money, classification
To bracket or not to bracket? p. 65 BLM 4.4, .45	8–13 year olds	Reading mathematical text	**Written language:** reading mathematical text	Brackets, addition subtraction, equations
Reading and writing division p. 66 BLM 4.6, 4.7	9–13 year olds	Reading and writing mathematical text	**Written language:** reading mathematical text Spoken language: explanation	Division, whole numbers, fractions

Monitoring writing in mathematics p. 68 BLM 4.8	All ages	Range of writing activities	**Written language:** writing	General
String activity p. 74	5–8 year olds	Writing in an activity	**Written language:** writing descriptions and explanations	Length, ordering shape
Writing instructions p. 79	8–13 year olds	Writing instructions	**Written language:** reading and writing instructions	Measurement of length

CHAPTER 5   GRAPHIC COMMUNICATION AND VISUAL REPRESENTATION

Exploring calculators p. 84	5–9 year olds	Drawing	**Graphic communication:** drawing Written language: writing	Calculator
Diagrams to solve problems p. 85	7–13 year olds	Using diagrams	**Graphic communication:** drawing diagrams Written language: reading and writing	General
Interpreting drawings p. 86	5–8 year olds	Interpreting drawings as 3D objects	**Graphic communication:** interpreting drawings, making models	2D and 3D shapes
Drawing a 3D object p. 86	8 year olds–adults	Drawing	**Graphic communication:** drawing	3D shapes
Using tables to solve problems p. 88	8–13 year olds	Making and using tables	**Graphic communication:** making tables Written language: reading and writing	Money, combinations logical thinking classification
Timetable problems p. 90	8–13 year olds	Interpreting timetables Posing problems	**Graphic communication:** interpreting tables Written language: writing	Time
General activities with tables p. 91	All ages	Using tables	**Graphic communication:** interpreting and making tables	
Find my house p. 92	9–13 year olds	Interpreting and using different co-ordinate systems	**Graphic communication:** interpreting maps	Maps, co-ordinates, decimals
Dot-to-dots p. 93 BLM 5.2	8–13 year olds	Using co-ordinates	**Graphic communication:** drawing and interpreting using co-ordinates	Co-ordinates, decimals
How many in a packet? p. 95	6–13 year olds	Collecting, organising and recording data	**Graphic communication:** making graphs, lists, tables	Graphs, chance, average

A country walk p. 98 BLM 5.3	10–13 year olds	Interpreting graphs	**Graphic communication:** interpreting and making graphs Active communication: physical involvement, measuring	Graphs, time distance
Who's who? p. 99 BLM 5.4, 5.5	8–13 year olds	Making and interpreting scattergrams	**Graphic communication:** interpreting and making graphs	Graphs, height, age, other measurements, correlation
Transferring information p.100	8–13 year olds		**Graphic communication:** interpreting tables, making graphs, drawing Written language: writing	
Graphing the times tables p. 101	9–13 year olds	Making a line graph	**Graphic communication:** making and interpreting graphs	Graphs, relationships interpolation
Draw my house plan p. 104	5 year olds–adults	Making a map	**Graphic communication:** drawing maps	Maps
A walk around the school p. 105	5–13 year olds	Making a map	**Graphic communication:** drawing maps Active communication: physical involvement	Maps
Walking to the shops p. 107	6–13 year olds	Making a map	**Graphic communication:** making maps	Maps
A map of the school grounds p. 108	6–13 year olds	Making a map	**Graphic communication:** making maps	Maps
Story map – Red Riding Hood p. 109	7–11 year olds	Making a story map	**Graphic communication:** making maps	Maps

1 | LANGUAGE MODES

Language and learning

Language and communication are essential elements in all learning, including mathematics. Children, from their earliest years, learn by actively investigating the world around them. They touch, feel, taste, observe, collect, pull apart and put together the things around them. However, active involvement in learning is more than simply *doing* these things. It also means thinking, reflecting, organising, and applying what has been learned to other situations. For example, as children compare size and quantity, colour and shape, they are doing more than observing and manipulating. They are making decisions, abstracting, and applying their knowledge to new situations. As they construct models, they are designing, predicting, and evaluating. Gradually, they make sense of the world around them. Language and communication play an essential part in this process.

As children continue to learn, the language elements are interwoven with the skills and processes of observing, comparing, sorting, classifying, making hypotheses, testing, designing, constructing, evaluating. Doing, talking, drawing and writing are ways of exploring and understanding the content of whatever is being learnt. Children bring their knowledge and understanding of the world around them to the classroom and this influences their understandings and beliefs.

If children's experience and knowledge are denied in the classroom, they may conclude that school knowledge is different, and separate, from what they learn out of school. We need to value what children bring to school, build on their experience, and challenge and extend their knowledge through further experience, discussion and reflection.

FIND OUT WHAT YOUR CHILDREN KNOW

All ages – adapt accordingly

Before starting a topic, find out what your children know about it. You can do this in a number of different ways.

- You can allow for free response, or focus the responses by asking questions such as:
  - How do people use X?
  - Where would you find X?
  - Think of as many different examples of X as you can.

BLM 1.1 can be used by individuals or small groups before whole-class discussion.

- Make a class list. Write on the blackboard or on a large sheet of paper

                    WHAT WE KNOW ABOUT …

Ask children to contribute to the list. Make sure that each child makes a contribution.

- Children can write individual statements.
- Children write a 'story' that uses the concept or skill. (See chapter 2 p. 14 for an example using decimals.)

See also chapters 3 and 4 for further suggestions that will help you to find out what your children know and think.

Language issues in the mathematics classroom

The words 'mathematics' and 'language' are combined in many different ways:

- mathematical vocabulary and terminology
- mathematics as a language
- acquiring the language of mathematics
- links between the language and the mathematics curriculum
- reading and writing mathematical language
- language as a factor in learning mathematics
- applying language learning principles to the teaching and learning of mathematics

In this book we will explore some of the complexities and varied meanings of the mathematics and language issues implicit in the above phrases. The focus is on:

- the language used in the mathematics classroom, and

- ways of using language in order to improve, extend and enrich mathematics learning and the understanding of mathematics.

Activities will be included which will provide opportunities:

- for you to explore and expand your mathematics practice in the context of language learning
- for children to expand their use of language in the mathematics classroom

---

GENERATING DISCUSSION ON MATHEMATICS AND LANGUAGE

An activity for groups of teachers or parents

- Ask participants to write down as many ways that they can in which mathematics is linked with language. For example, 'mathematics as a special language'.
- In small groups, they compare their lists and clarify what they mean by each item.
- They then write a new 'agreed' list that includes all their ideas.
- Display these as a whole group, and compare the lists. Discuss and clarify meanings. Discuss:
  - Which of these do we currently include in our mathematics programs?
  - Which of these should we put more emphasis on in the future?

As a result of the activity, the complexity of the issue will be recognised. There will be some agreed and shared understandings among the staff and parent body, which are essential before moving into curriculum development and curriculum change.

Communication modes

People communicate in many different ways. The most obvious is through language, both spoken and written. People also communicate through graphics – drawings, diagrams, graphs – and through their actions – demonstrating, performing, body language. For conciseness, we use the word 'language' in this book to refer to all these forms of communication, not only to spoken and written language.

The aim of the book is to show how the four major communication modes can be used within mathematics learning to improve and extend students' learning. These modes can be categorised as:

- spoken language (speaking and listening)
- written language (reading and writing)
- graphic representation (diagrams, pictures, graphs)
- the 'active' mode (performing, demonstrating and physical involvement)

These are all important in effective mathematics teaching and learning, and work to extend children's mathematical experiences and perceptions.

Each mode has two aspects:

- receptive (processing someone else's communication)
- expressive (using your own language)

Del Campo and Clements (1987) point out that, traditionally, mathematics classrooms have emphasised receptive language; expressive language has largely been limited to copying forms and procedures demonstrated by the teacher. Table 1.1 has been adapted from Del Campo and Clements.

Table 1.1: Communication modes

Mode	Receptive	Expressive
Spoken	Listening	Speaking
Written	Reading	Writing
Graphic	Reading and interpreting diagrams and pictures	Creating and drawing
Active	Interpreting others' actions	Performing, demonstrating, physical involvement

Not only have the receptive modes been emphasised to the neglect of the expressive modes, the forms of language to which children have been exposed have also been limited. In particular, they have been limited to teacher- and text-directed forms:

- listening to the teacher
- answering closed questions
- reading 'sums' from the blackboard or textbook
- writing algorithms

In the traditional mathematics classroom, students were required to work silently and individually, reading was confined to the textbook, worksheet and blackboard, and writing in mathematics involved only symbolic statements (such as $3 + 5 = 8$) or writing algorithms. Graphic communication was restricted to the formal reading and making of graphs, and active communication was limited to replicating the manipulation of concrete materials as demonstrated by the teacher.

Del Campo and Clements argue that classrooms need to encourage expressive language — and not only in the form of discussion. The whole range of communication modes – oral, written, graphic and active – should be used, in order to broaden and deepen students' understanding of mathematics.

The traditional approach largely ignores the expressive modes of communication and has denied children opportunities to investigate mathematics and to assimilate knowledge into meaningful contexts. Such an approach has resulted in a narrowing of the mathematical concepts to those demonstrated or modelled by the teacher or the textbook: an example given, then practice by students.

The outcome for the curriculum is often an emphasis on symbols, terminology and manipulation rather than on concepts, meanings and applications. The outcome for students is the belief that mathematics consists of symbols, terminology and manipulation, with little purpose or relevance.

The following tables outline how each of the communication modes can be expanded, and provide indications of how to extend the use of language in the mathematics classroom.

Table 1.2: Spoken language

	RECEPTIVE Listening	**EXPRESSIVE** Speaking
Traditional practice	to the teacher	answering closed questions
Expanded practice	to each other to parents	to each other to parents to other members of the school community discussing open-ended problems drama and role play explaining and justifying questioning others presenting reports

Table 1.3: Active communication

	RECEPTIVE Interpreting	**EXPRESSIVE** Performing and demonstrating
Traditional practice	interpreting the teacher's actions	manipulating materials
Expanded practice	watching and interpreting the actions of other students	drama role play mime demonstrating physical activity

Table 1.4: Written language

	RECEPTIVE Reading	EXPRESSIVE Writing
Traditional practice	sums word problems	'fill the gap' 5 + 7 = algorithms

Expanded practice	stories
	
	non-fiction
	
	newspapers and magazines
	
	each other's writing
		factual, creative and imaginative
		  descriptions
		  recounts
		  explanations
		  reflection
		  stories
		  journals

Table 1.5: Graphic communication

	RECEPTIVE Reading & interpreting	EXPRESSIVE Creating & drawing
Traditional practice	graphs and diagrams presented by teacher or textbook	graphs and diagrams under direction
Expanded practice	each other's work graphs and diagrams from books, magazines, newspapers maps	recording in own ways representing with concrete materials drawing maps creating graphs and diagrams from raw data

Expanding the traditional language modes provides a variety of language contexts that will result in a rich learning environment for both teachers and children. Changing the emphasis from predominantly a receptive language to a more balanced mix of receptive and expressive language provides opportunities for children to be actively involved in their learning, and to gain a deeper understanding of mathematics. By expanding the language modes of the classroom, teachers give children opportunities to:

- create their own understandings and interpretations
- clarify understandings
- address misconceptions
- work in a variety of ways
- develop skills using a variety of learning strategies

- use preferred learning strategies
- develop skills in using other learning strategies
- consolidate knowledge
- build on previous knowledge
- share their understandings with others and so broaden their understandings and applications

Teachers are enabled to:

- gain further insights into the children's development of mathematics concepts and understandings
- provide evidence of children's progress to report to parents
- plan more effectively by building on what children know

The language modes do not occur independently of one another; rather, there is an extensive overlap. For example:

- during discussion, children may be making notes or drawing diagrams
- when presenting a talk, written and graphical material may be used as illustration
- when writing, whether informal notes on a problem solution, or a poster to be presented for a project or investigation, there is likely to be a mixture of written words and sentences, symbolic or algorithmic mathematical statements, and diagrams, pictures or graphs

The teacher's language

As a teacher your language should communicate:

- general expectations of children: for example, that they will respond to a challenge, that they will persevere, and that they are able to succeed;
- the expectation that all children, regardless of gender, language, cultural background or social class, can learn mathematics;
- values and beliefs about learning, about mathematics, and about learning mathematics. For example, that mathematics is relevant and useful, enjoyable and challenging;
- clear expectations of specific tasks. For example, that you expect an explanation of the method, and a justification of the solution, not only an answer, when solving a problem;
- that you value children's contributions and knowledge.

As well, you need to model, demonstrate and explain:

- appropriate mathematical vocabulary, syntax and symbols
- the language of explanation, hypothesis and justification
- ways of questioning, discussing, and exploring mathematical ideas
- methods and strategies for solving mathematical problems

There is much to attend to in the classroom, and it is easy to overlook the impact our language may be having on children's attitudes, self-esteem and achievements.

Modelling ways of talking about mathematics

By accepting children's own language, and, at the same time, modelling and explaining mathematical language, you will help children to extend their own use of language. As well as using terminology and symbols correctly and in contexts that make sense to children, you can also model and demonstrate the processes of mathematics which are expressed in language. You can use the following strategies:

- **Model distinctive language forms and vocabulary.**

  For example, you can use technical vocabulary in context, together with everyday language. You can use constructions such as 'I think that ... because ...', or 'I wonder if ...', and qualifiers such as 'probably', 'perhaps', 'always'. You can discuss with the children why and when these language patterns are appropriate.

- **Ask 'good questions', and encourage children to ask themselves these kinds of questions.**

  'Good questions' require more than the recall of a fact or the reproduction of a skill. They extend the child's learning and your knowledge of the child, and they may have several acceptable answers (Sullivan & Clarke 1991). Specific examples of such open ended questions in mathematics are given by Sullivan and Clarke (e.g. p.37).

POSING GOOD QUESTIONS

All levels, all ages (Adapt according to the ages of your group.)

Construct 'good questions' in any topic by one of these methods.

Method 1. Working backwards

1. Identify a topic
2. Think of an answer
3. Make up a question that includes the answer

For example, in the topic 'addition of fractions', the answer might be $1\frac{1}{2}$. The question could be:

  If Jo and Camilla ate $1\frac{1}{2}$ pizzas between them, how much might each of them have eaten? Is there more than one answer?

Method 2. Adapting a standard question

1. Identify the topic.
2. Think of a standard question or task.
3. Adapt it to make a 'good' question.

For example, in the topic of measurement of area, a standard question might be:
Find the area of this triangle.
A good question might be:
Draw a different triangle with the same area as this one.

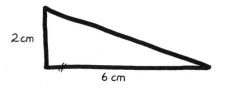
2 cm
6 cm

Adapted from Sullivan & Clarke, 1991, pp. 20–1

- **Ask the children general questions relating to specific topics or information such as:**
  - What do we know?
  - What do we want to find out?
  - Is it true?
  - How do we know it is true?
  - Is it always true, or could it change?
  - Do you agree? Why, or why not?
  - What would happen if?
  - Would that work if ?

Encourage the children to use questions like these themselves when they are working. List and display the questions and refer the children to the list when they are working on a task. Give the children positive feedback when they use these questions.

- **Model the ways questions could be answered.**

For example, you can explain how you guessed the number of jelly beans in a jar, and then ask students to explain their methods:

> 'I counted 25 I could see on the bottom layer. Then I counted 12 up the side to see how many layers there were. And then I multiplied 25 by 12, and got 300.'

- **Make the language explicit by discussing and listing the kinds of questions which help us to think about and understand the mathematics we are using.**
- **Give feedback when children ask appropriate questions or provide clear explanations:**

> 'That was a good question.'

> 'Your explanation was very clear.'

> 'I noticed that you thought of a different method from the other groups.'

Activities for students focusing on the development of mathematical language, its vocabulary, syntax and recording are discussed in the next chapter.

2 | DEVELOPING THE LANGUAGE OF MATHEMATICS

If children are to develop and effectively use the language of mathematics, we need to help children to understand the special features of mathematical language – vocabulary, syntax and symbols – and at the same time we need to demonstrate that mathematics is useful, relevant, interesting, and a part of everyday experience. Unless we can establish first that mathematics is useful and meaningful, we have little hope of showing that mathematical language is useful and meaningful.

The language of mathematics

The language used in mathematics, and in particular the language encountered in mathematics classrooms, is distinctly different from the everyday language of children and adults. Every discipline has its own register, which 'experts' have mastered and 'novices' need to learn, but mathematics appears to be more difficult than most to master, for two reasons.

The first is the nature of the mathematics register. It is a dense and precise language. Every word or symbol is important to the meaning. For example, the use of brackets in mathematics usually changes the meaning of a mathematical statement, where in 'normal' writing brackets indicate that the bracketed phrase or clause adds additional information, or illustrates or clarifies the main text:

$$(10 - (4 + 2) \neq 10 - 4 + 2)$$

Reading mathematics is discussed further in chapter 4.

The second reason relates to the way we have traditionally taught mathematics. We have ensured that many children see mathematics as the manipulation of symbols and formulas, and the language of mathematics as difficult, irrelevant and meaningless, by practices such as:

• introducing vocabulary without any meaningful context

• insisting on 'correct' setting out

• emphasising symbolic mathematical language

• restricting opportunities for children to explore both the mathematical concepts and the ways they can be expressed

There is strong evidence that requiring children to write mathematics in formal and conventional ways does not necessarily mean that they will use these methods in other situations. We have found, in many classroom sessions where children had to solve the particular problem of sharing a large number of cookies among twelve children, that very few will write a division equation or division algorithm in the conventional forms (e.g. $72 \div 12 = 6$, or $6\overline{)72}$), although they have used these forms extensively in other class lessons.

Martin Hughes (1986 pp. 55, 72–5) had a similar finding for the use of the symbols +, −, and =. Children whose workbooks contained pages of sums such as $7 + 5 = 12$ and $8 - 3 = 5$ did not use these symbols when asked to show what happened when a number of bricks were put together or removed. Nor did they use + and − symbols when asked to send each other written messages about the number of bricks to add to or take off a tower.

The following activity involves children in recording a number of blocks and changes to that number so that another child can 'copy' the first child's tower.

NUMBER MESSAGES

Years K–2, ages 5–8

• Children work in pairs, with a screen between them.

One child builds a tower with Unifix or other blocks, up to ten blocks high.

S/he sends a message on paper to the other child so that s/he can build one the same height.

Then the teacher takes a number of blocks off (or adds onto) one child's tower.

The child sends a message on paper to tell his/her partner how many to take off (or how many to add on).

• Display the children's work.

• Ask children to explain their recordings.

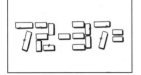

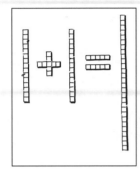

In some situations, the symbolic form takes on a life of its own. We recall the year 4 children who were having difficulty solving 72 – 37. They were asked, 'Could you use MAB blocks to work it out?' They used the blocks to 'write' the problem in numerals, not to represent the numbers!

A similar case occurred with two year 1 boys, who wanted to extend the addition equations they had been writing to larger numbers. They wrote 15 + 15 = 28. Their teacher challenged them to show this with the Unifix, so they used the Unifix as shown at left. In this case, they are very close to the correct answer – but instead of using the Unifix to check their solution, they have used the material to present their answer.

Representing mathematical ideas and problems

When children or adults collect or manipulate information (or data) they can communicate or represent this in many different ways. They may give an oral description, which may or may not include information about number, size or shape. They may draw a diagram or a picture, record using tallies, or use numerals and other mathematical symbols. They may write in a range of forms. They may draw a graph of some kind.

Whatever method is used, each representation shows only some aspects of the situation. For example, if children are given the problem of sharing 20 Smarties among 5 children, there are a number of ways they can represent the problem in order to solve it.

- They may physically share out real Smarties among 5 children, dealing them out one by one.
- They may use counters to represent the Smarties, and deal the counters into 5 piles.
- They may draw a picture of the five children, with four Smarties beside each child.
- They may write an equation 20 ÷ 5 = 4.
- An older student may generalise the situation further, by writing an algebraic relationship:

$$s = S \div n$$

where s is the number of Smarties each child gets, S is the total number of Smarties and n is the number of children.

Each of these methods enables different aspects of the problem to be addressed. To the adult who has proposed the question, the number of Smarties each child receives is likely to be the only aspect to be considered. For some children, other aspects such as the colour of the Smarties, and whether they or their friends like chocolate may be more important.

- Sharing real Smarties will enable aspects such as colour, and individual preferences, as well as fairness, to be addressed. But the actual number each child receives may never be shown. There is no need to count, and anyway the children may have eaten some before the sharing is complete.

- Dealing counters into five piles will make the equal size of each pile easier to see and to count.

- Drawing a picture will enable children to ensure that their favourite colours are included in the selection. Some counting will be necessary to ensure that there are still 20 Smarties after the sharing.

- The equation $20 \div 5 = 4$ will tell how many each child gets. It does not tell what colour each child gets. So it shows less information than the drawing but, on the other hand, is more efficient if all we want to know is the number to give each child. It can also be applied to many different sharing situations, where the drawing is specific to the particular situation.

- The equation $s = S \div n$ has moved from the specific situation to a very general statement. Although it can be used in a range of situations, it cannot be applied in a rote fashion to any particular situation. For example, if S is not exactly divisible by n, can s be a fraction or decimal?

REPRESENTING DIVISION

Years K–4, ages 5–10
- Pose the problem:
  How could you share 20 Smarties among 5 children?
  (Choose appropriate numbers for the group of children.)
- Provide a range of material, including coloured pencils, paper, counters, sticks, bead frame or abacus, calculators.
- Ask the children to work out the answer using two different methods.
- Display and discuss the representations.
- Ask children to explain what they have done.

DIVISION STORIES

Year 5 up, age 10 up

- Ask children to write about situations for each of the equations below, where the given answer would be appropriate.

  $20 \div 3 = 6$

  $20 \div 3 = 6$ remainder 2

  $20 \div 3 = 7$

  $20 \div 3 = 6^{2}/_{3}$

  $20 \div 3 = 6.7$

  $20 \div 3 = 6.6$

- Display children's work.
- Look for similarities in children's recordings.
- Ask children to explain their 'stories', especially if they are 'bizarre' or different.
- Use the information from this activity to plan other activities related to division.

Assessment

- Do children find appropriate situations for each answer?
- Note the answers that children find difficult to account for, and plan activities focusing on appropriate situations.

The following activity focuses on identifying appropriate uses and applications for decimals.

DECIMAL STORIES

Year 4–adult, age 9 up

- Write up the equation

    $4.6 + 5.3 = 9.9$

- Ask children to write a story to match the equation.
- Ask children to explain their responses, and allow others to challenge and question.

Assessment

Children's responses will give you a quick assessment of their understanding of decimals.

Use children's responses as discussion starters to explore the meaning and uses of decimals.

This kind of activity contributes to building a classroom climate of open discussion and trust, while at the same time it has a strong mathematical focus.

Mathematics provides concise and powerful ways of communicating information and predicting results. But it cannot tell everything about a situation.

Because any representation tells only part of the story, it is important to use a range of representations, and to allow children to experiment with different forms. They can

- represent the aspects which are important to them
- compare their methods with others
- learn different aspects of recording information from each other

The freedom to record in their own way, followed by sharing, explaining, and discussion, provides an opportunity for learning about effective and clear methods of organising and presenting information.

FORMAL AND INVENTED NOTATION

Inventing notation is part of developing new mathematical ideas. Some notations have proved to be less powerful than others. For example, the notation developed by Isaac Newton for the calculus was discarded in favour of the notation developed by Leibniz. While still at school, Richard Feynman (who later won a Nobel prize for physics) invented his own symbols for trigonometry and for calculus, which in his view were superior to and more logical than the conventional notation (Feynman 1986). However, he realised that a notation which cannot be used to communicate with others is of limited value, and reluctantly returned to the conventional notation.

So children who invent their own notation are in good company. They are thinking mathematically. If children are allowed to and are actively encouraged to record and represent mathematical information in their own ways, they will do so in ways which make sense to them. They will also better understand the uses of the formal language when it is introduced. Indeed, often they will perceive a need for words or symbols to describe some process or concept, and will either invent their own symbol or ask for the 'correct' method of recording.

James, five years old, was playing a game based on the book *The Python and the Pepperpot*, in which a python 'eats' a number of children (represented by counters). The child has to say how many of the six children have been eaten by counting the number left. The children had played the game several times, and were asked this time to record the result of each game. James wrote

$$6 \, \text{t} \, 9 \, 2 = 4$$

and read this as 'six take away two equals four'. James had invented his own notation for this process, as he perceived a need and a purpose for the notation.

In a year 1 class, children were given 10 beads or counters. They were asked to give away some to a partner and record what happened to the numbers. They recorded in a number of ways. After a while, one pair asked the teacher for the 'proper way' to write take away. Like James, they saw a need for the notation.

Jane and Kathryn wrote 10 at the right side of the page, and took away 1 (top line), 2 (second line), and 3 (bottom line).

Mark and Tim made a table. Their counting was correct, although they lined the numbers up incorrectly.

THE WOLF GAME

Years K–2, ages 5–8

• Read the story 'The Wolf and the Seven Little Kids' (see BLM 2.1) to the children, and discuss what is happening to the number of kids as the wolf swallows them.

Alternatively, use the book *The Python and the Pepperpot,* or another story in which a number of things, people or animals are eaten one by one and then reappear (e.g. *The Fat Cat*).

• Role play the story with the mother goat, the seven kids, and the wolf in front of the class. As the wolf 'eats' each kid, the kids move behind the wolf, so that the class can see easily how many have been eaten and how many are still to be eaten.

• Introduce the game, which children can play in pairs. Each pair needs seven counters (or some other agreed number) to represent the kids. The wolf can be represented by a sock over one child's hand. The 'wolf' picks up and conceals some of the 'kids'. The other child has to say how many have been 'eaten' by seeing how many are left.

• When children have played the game a few times, and are familiar with it, ask them to record what happens to the numbers each time. How many are there to start with, how many does the wolf eat, and how many are left? Allow them to work out their own method for recording.

Assessment

· Ask children to explain how they worked out the number eaten.

· Ask children to explain their recordings.

· Note children who are having difficulty with the concept of subtraction.

· Give children the opportunity after discussion to repeat the activity, in this context and in other contexts, for example the *Cups Game* on BLM 3.3.

Other activities

· List the positional vocabulary used — *in, under, between,* etc. Act out and draw pictures to show their meanings.

· Draw a plan of the wolf's house to show where each kid hid.

BLM 2.1

THE WOLF AND THE SEVEN LITTLE KIDS

A mother goat lived in a house with her seven little kids. She knew a hungry wolf lived nearby so she looked after the kids every minute of the day.

One day the mother goat had to go out. She warned her children not to open the door to anyone except herself, and especially not to the big bad wolf.

'How will we know if it's the wolf?' asked the eldest of the kids.

'You will know him by his deep growly voice and his black paws,' said the mother goat. 'My voice is high and sweet, and my feet are white.'

When the mother goat had left the house, the wolf came along and knocked at the door.

'Let me in,' he growled.

'Oh no,' said the second little kid. 'We can only let in our mother. She has a sweet, high voice, not a deep, growly voice. You are the big bad wolf.'

So the wolf went away, angry.

He went to a beekeeper who lived nearby, and said 'Give me some honey, or I will eat you up.'

The beekeeper gave the wolf a large jar of honey, which the wolf gobbled down. It made his voice sweet and high. So he returned to the goats' house.

'Let me in,' he said, in his new sweet, high voice.

The third little kid was about to open the door, when the fourth little kid said 'Wait! Show us your feet.'

So the wolf put his feet up on the windowsill, where the little kids could just see them.

'Oh no!' said the fifth little kid. 'You are not our mother. You have black feet, and she has white feet. You are the big bad wolf.'

So the wolf went away again, angrier still.

He went to the flour mill, and said to the miller 'Give me some white flour, or I will eat you up.'

The miller gave the wolf a bag of flour, which the wolf used to dust his feet, and turn them white, before he returned to the goats' house.

'Let me in,' he said, in his sweet, high voice. 'I am your mother. See my nice white feet.'

He put his feet up on the windowsill to show the little kids.

'Yes, I'll let you in mother,' said the sixth little kid, and opened the door.

BLM 2.1 (cont)

What a common_____ chased them!

He caught the fi_____ him whole.

He caught the s_____ her whole.

He caught the th_____ swallowed him wh_____

He caught the fo_____ her whole.

He caught the fi_____ her whole.

He caught the s_____

The seventh littl_____ not find her.

So the wolf stag_____

Soon the mothe_____ the house empty.

She called for he_____ grandfather clock._____ brothers and sister_____

The mother goa_____ off with the sevent_____

Soon they caugh_____ under a big tree.

The mother goat took out her scissors, and cut a slit in the wolf's stomach. The wolf was so fast asleep after his greedy meal that he did not wake up.

Out jumped the sixth little kid, alive and well.

Out jumped the fifth little kid, alive and well.

Out jumped the fourth little kid, alive and well.

Out jumped the third little kid, alive and well.

Out jumped the second little kid, alive and well.

Out jumped the first little kid, alive and well.

'Fetch me a large stone,' said the mother goat, and the seventh little kid ran quickly to find one.

The mother goat placed the stone in the hole she had cut in the wolf's stomach, and sewed up the slit.

Then the goats all ran home and never opened the door again when their mother was out.

VARIETY OF REPRESENTATIONS

Year 3 children had read the book *A Bag Full of Pups,* in which Mr Mullins gives away twelve puppies to a range of homes. The children were asked to represent what was happening to the number of puppies at some stage in the story. Equations, number lines, diagrams and drawings were all suggested. One boy focused on the number and drawings were all suggested. One boy focused on the number twelve, and had two unusual suggestions.

'If you had one of those old rulers that goes up to twelve, you could cut off one mark for each puppy he gives away,' and

'Start with the clock at twelve o'clock and go back one hour for each puppy.'

These may not be the most efficient or clearest ways of representing the situation. But they are legitimate and logical models for demonstrating subtraction, and focus on the total number of puppies — twelve — by using models which go up to exactly twelve. Being able to find analogies and to make connections is both creative and useful.

HOMES FOR ANIMALS

Years 1–3, ages 6–9
- Children will be familiar with the problem of finding homes for litters of baby animals, whether puppies, kittens, rabbits or guinea pigs.
- Discuss with the children their experiences, or the experiences of friends, neighbours and relatives. If available, read *A Bag Full of Pups*.

This will be an opportunity to discuss and emphasise the social responsibility involved in keeping pets.
- Ask children to make up a story about finding homes for a litter of baby animals.
- Ask the children to represent what happens to the numbers each time they find a home for one of the animals. They may draw, write, or use concrete materials for their representations.
- In pairs, the children read their stories to each other.
- In groups of four, the children choose one to read to the rest of the class.
- Display the children's recordings and representations.
- With the children, write statements about the recordings.

Assessment
- Note the range of recording methods used.
- Are the recordings clear?
- Note information which will be of use in planning further activities.

VARIETY OF MATERIALS OR MEDIA

Recording can be done using a range of materials and media. In the following activity, calculators, concrete materials and paper were all used as recording tools. A group of 5- and 6-year-old children were making number patterns on calculators. After a while they were asked to make one of their patterns on the calculator and with counters, Unifix, sticks or other concrete material. We had expected them to make the pattern, for example 12121212, by putting out a single counter, then a group of 2, 1 then 2 and so on.

Instead, many of the children put out counters of two different colours, for example one red counter, followed by one yellow counter, followed by one red and so on. One child put out a row of random colours to represent 78787878, and explained that the first counter was for the first 7, the next for the 8 and so on. When asked how she could tell which was a 7 and which an 8, she didn't see any need to differentiate the 7s from the 8s. However, later on we noted that she had offset the counters so that the 7s were above the 8s.

Other children also modified their patterns as they saw what other children had done, or after explaining their patterns to each other or to one of the adults in the room. Quite a few now used the number of counters to correspond with the numerals they had made.

The children were then asked to record their patterns on paper. They did this in a variety of ways, some copying the numerals on their calculators, some drawing the patterns they had made with counters, and some both. Many extended their patterns to cover the whole line or the whole page.

What did the children learn from this activity?

- They learnt something about the enormous variety of number patterns which can be created.
- They learnt that these number patterns can be transferred to other contexts or media, and that this can be done in a number of different ways.
- They learnt which materials were more useful for certain aspects of the task.
- Many children developed new ways of creating patterns, both from their own explorations and from observing and discussing with others.
- They also learnt that they were able to create and manipulate patterns successfully.

What did the teacher learn from the activity, to help in assessment and planning?

- The teacher was able to identify children who had not yet developed ideas of pattern and those who were confident and competent in creating patterns.
- She was able to identify those children who interpreted numerals as representing a number of items, and those who saw the numerals as single items.
- By looking at their recordings, she was able to identify children who could not only count a number of items correctly, but could draw that number correctly. When drawing a stick of Unifix or a pile of blocks, this is not so easy.
- She was also able to identify a number of children whose understanding appeared to develop during the session.
- By listening to children's explanations of their patterns and their recordings, she gained further information on individuals' understandings of numbers and of patterns. She had recorded information pertinent to the children's mathematical development.
- By the end of the session she had developed plans of her own for further work with calculators and with patterns.

CALCULATOR PATTERNS

Years K–3, ages 5–9

- Provide one calculator between every pair of children. Explain that they need to take turns using the calculator, and that the calculator must stay where they can both read the display.

If the children have not used calculators before, give them time for free exploration, before starting any structured activities.

- Show the children a number pattern on blackboard or poster, for example

12121212

- Ask the children to make the pattern on the calculator.
- Give the children practice in making different patterns, for example:

57857857   87654321   24680246   48484848

- Now ask each pair to choose one of the patterns they have made, display it on the calculator, and make the same pattern using concrete material.
- When they have done this, ask them to draw the pattern on paper.
- Finally, the children can write about the activity.

Assessment

- Note children who interpret the numerals as marks, rather than as numbers.
- Are children able to copy the patterns accurately on the calculator?
- Can children represent the pattern satisfactorily in concrete materials?
- Were children able to draw the patterns correctly?
- Could children develop their own patterns?
- Note additional information provided in the children's writing.

We have to be aware that, just as these young children did not use the concrete materials in the way the teacher expected, many older children will also use materials in (to us) unexpected ways.

Developing the language of mathematics

As stated earlier, the mathematics register (vocabulary, syntax, symbols and structures) poses more difficulty to the novice than that of many other disciplines. Some ideas and suggestions for developing the language of mathematics are discussed here.

MATHEMATICS VOCABULARY

The particular vocabulary and terminology of mathematics have long been identified as posing specific problems to children's understanding. Words such as *area, volume, right, odd, difference, factor, diagonal* all have specific mathematical meanings, as well as commonly used everyday meanings.

It is important that the uses of words such as these be explored explicitly in different contexts, for example by asking children to make up 'riddles' for words with more than one meaning. One such riddle would be:

It tells you how loud it is.
It also tells you how much milk there is in the carton.
It can also be a book.
What is it?

Other words, such as *perpendicular*, *bisect*, *divisor* and *quotient* are not commonly used outside the mathematical context. Their particular meanings will need to be taught, as with terms specific to any curriculum area, both by modelling the use of the words in context, and of course by explicitly providing definitions and examples.

Children can be helped to understand and use these words correctly through a range of activities, for example:

- creating a class maths dictionary. This can be an ongoing project throughout the year;
- making a glossary. This can be in relation to a specific topic;
- writing an explanation for another child;
- discussing the roots of words;
- listing multiple meanings.

MAKE A BOOK OF MATHEMATICAL RIDDLES

Years 2–7, ages 7–13

- With the children, list words which are used in mathematics, and which also have an everyday meaning. Some words to start you off are given below, however, it is best to work with those words suggested by the children, and those you are using regularly in the classroom.

acute	difference	mean	remainder
alternate	digit	net	right
angle	dividend	odd	scale
area	even	operation	section
chord	face	oval	similar
degree	factor	power	translation
diagonal	figure	product	volume
diamond	index	reduce	

Discuss the meanings of the words. Ask children to explain the meaning in their own words.

- In pairs, children choose a word to write a riddle about.
- When drafts are completed, and have been checked for accuracy, the children write their riddle on one side of a page. On the other side, they write the answer and draw a picture showing both the mathematical and everyday meaning.
- Make the riddles into a book, which children can borrow to read and take home. They can also read the riddles to children in other classes.

BOOK OF NUMBER WORDS

Years 3–7, ages 8–13
(Adapt the activity as appropriate for your group of children.)
• Compile a book in small groups. Each group takes one (or more than one) number between 1 and 10. You may wish to include other numbers such as 12, 20, 100 and 1000.
• List all the words which include that number in their meaning, e.g. monocycle (one wheeler), hexagon (six-sided figure), heptathlon (sport with seven different events). Some suggestions to help children get started are given on BLM 2.2.
• Each group designs and makes a page for their number, which includes all the words, their meanings, their origins, and illustrations.

Assessment
• Are children able to identify words which are linked to particular numbers?
• Can children explain clearly the meanings of the words they find?
• Useful resources: an etymological dictionary (one which gives word origins), *The Guinness Book of Numbers*, a mathematics dictionary

BLM 2.2
NUMBER WORDS
Many words relating to number come from Latin and Greek number words. Look for the following elements:

	Latin	Greek	Other
1	uni- as in unicorn	mono- as in monocycle	words based on one, such as alone, once, only
			some words for first starting with pri-, such as prince, prime
2	bi- as in biplane du- as in duet	di- as in dilemma	words based on two, such as twice, between, twig
3	tri- as in triangle	also tri- as in tripod	
4	quadra- as in quadrangle, quart- as in quartet	tetra- as in tetrahedron	
5	quintus- as in quintuplet	penta- as in pentagon	
6	sexta- as in sextant	hexa- as in hexameter	
7	septem- as in September	hepta- as in heptathlon	
8	oct- as in octave	octo- as in octopus	
9	novem- as in November	nonus- as in nonagon	
10	deci- as in decimal	deca- as in decathlon	words such as dime, tithe, meaning one tenth
100	centi- as in centigrade	hecto- as in hectare	
1000	milli- as in millennium	kilo- as in kilometre	

TOPIC GLOSSARY

All levels, all ages
• When working in a topic, whether a mathematical topic such as graphs or an across curriculum topic such as the environment, compile a word list as new or unfamiliar mathematical terms are encountered.
• Write the words and their meanings, together with diagrams, on a wall chart, which children can refer to as they work.
• At the completion of the topic, children can sort the words into alphabetical order, and compile a class glossary. They may add additional information or refine and clarify the meanings at this stage.

Assessment
• Are children able to explain the meanings in their own words?
• Note whether children use the words listed spontaneously in subsequent work.

Another area of difficulty for young children is the use of comparative terms such as *taller, bigger, shorter* and so on. Intuitively, all of us, when we hear a person we do not know described as 'taller' than another, imagine a person above average height. When interpreting statements such as 'Jo is taller than Kim', many children infer that Jo is tall (because described as taller), and some also infer that Kim is short (in contrast to Jo). If an additional statement such as 'Kim is taller than Peter' is given, many children are confused because Kim appears to be both short and tall. (Donaldson 1963)

THE FOUR CROCODILES

Years 1–4, ages 6–10
- Tell of the four crocodiles named Sweetheart, Dundee, Snap and Hogan, who lived in the Daintree River.
  - Dundee was longer than Sweetheart.
  - Snap was shorter than Sweetheart.
  - Hogan was the longest crocodile.
- Ask the children to draw a picture of the crocodiles, showing them in order of size.

Assessment
- Are children able to list the crocodiles in the correct order?
- Do the children's pictures show the relative size of the crocodiles?
- Can children explain what they have done?

Children will need many experiences with physical materials and the concepts of tall/short, long/short, heavy/light, thick/thin, wide/narrow, ordering objects, describing the objects and how they are ordered in comparative terms. For example, when walking, is a long distance the same as when driving? Is a cheap house likely to be more expensive than an expensive car.

WHAT ORDER?

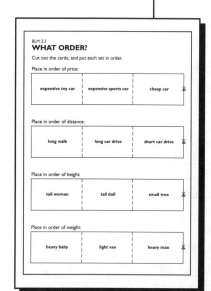

BLM 2.3
WHAT ORDER?
Cut out the cards, and put each set in order.

Place in order of price:

| expensive toy car | expensive sports car | cheap car |

Place in order of distance:

| long walk | long car drive | short car drive |

Place in order of height:

| tall woman | tall doll | small tree |

Place in order of weight:

| heavy baby | light van | heavy man |

Years 1–4, ages 6–10
- Cut out the cards on BLM 2.3 and give each set to a child or a pair of children to place in order.
- Ask the children why they have put them in that order.
- Children can write about what they have done, explaining the order.

Assessment
- Are children able to place the cards in order?
- Are some of the sets more difficult for children?
- Are children's explanations clear?
- Place in order of price: expensive toy car, expensive sports car, cheap car
- Place in order of distance: long walk, long car drive, short car drive
- Place in order of height: tall woman, tall doll, small tree
- Place in order of weight: heavy baby, light van, heavy man

Logical connectives such as *and, or, if, whenever, because, therefore* can also be problematic for children. These common words perform a crucial function in mathematical reasoning. Their use needs to be addressed explicitly, at appropriate times. Venn diagrams can help children to see the difference between *and* and *or*.

VENN DIAGRAM

Years 1–5, ages 6–11

- Make a Venn diagram on the asphalt with the children. First decide the categories that you will use. For example, boys and girls as one pair of categories, and two hobbies, sports or interests which are popular with the children, such as swimming, athletics, dancing, stamp collecting, tennis, for the second pair.
- Mark out an area on the asphalt, similar to the diagram below, making sure that the different areas are large enough for a group of children to stand in.
- Ask a number of children (up to about ten) to place themselves correctly, one by one, in the diagram.
- Repeat the activity with other groups until the whole class has both taken part and observed the activity.
- Now use the Venn diagram on BLM 2.4.

From the Venn diagram, ask the children:

- what you could call the whole set within the rectangle
- to list children who are boys or girls (all)
- why don't the rings for the boys and for the girls overlap?
- why do the rings for the children who play tennis and for the children who play basketball overlap?
- to list children who play basketball *and* tennis
- to list children who play *either* basketball *or* tennis

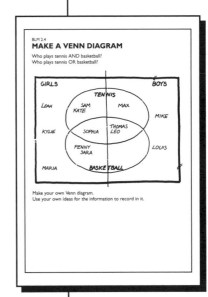

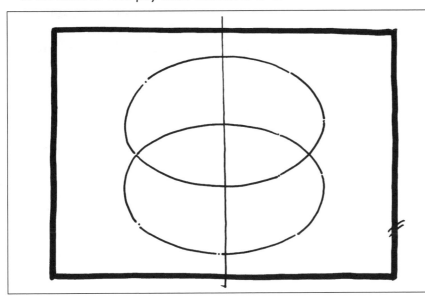

- Ask puzzles such as:

  I am a girl and I play basketball. I don't play tennis. My name has four letters. Who am I?

- Now divide the class into groups of between eight and twelve children, containing both boys and girls, and collect similar data for each group. They can choose their own categories. Each group can make a Venn diagram. Children can do this physically, and then record what they have done on paper.

Assessment

- Are children able to place themselves correctly in the Venn diagram?
- Can children distinguish between *and* and *or* using the diagram?
- Can children construct their own Venn diagrams?
- Can children interpret the Venn diagrams?

Second language learners

For many of the children in our classrooms, English is their second language. Bilingual children may have an advantage in learning and understanding mathematics: they experience concepts and skills through two languages and the two cultures belonging to these languages, which broadens their understanding. However, in the process of learning mathematics at school and, at the same time, learning competence in the language of instruction, they may need extra support.

These children need opportunities to answer questions in their first language, both orally and in writing, as well as in English. It may be possible to use other children of the same first language as mentors and as interpreters. Use demonstrations as well as oral and written instructions. Give students the opportunity to respond in practical ways, using concrete and visual materials, diagrams and so on, as well as in oral and written language.

Find out the numerals and number words used in the child's first language. (This is an excellent way to involve families in the mathematics learning.) Many languages make the place value system more explicit in the number naming than does English, particularly the notorious 'teen' numbers. For example, in many Asian languages the word for 16 is made up of ten-six in that order. Other languages show vestiges of a system of counting by twenties, for example French, where 80 is 'quatre-vingt', or 'four-twenty'. A chart with the numerals and number words written in English and in their first language will help many children. And exploring alternative numerals or number words, as a class or in small groups, will help all children to understand our own numbering system better, as well as showing that the knowledge each child brings from their own culture is valued.

NUMBERS IN OTHER LANGUAGES

Years 1–7, ages 6–13

(Choose the activities suitable for your age group.)

Ask children who speak a language other than English at home to bring to school information about numbers in their other language. BLM 2.5 can be used.

Use the information children bring as individual resources, and also as a resource for class lessons.

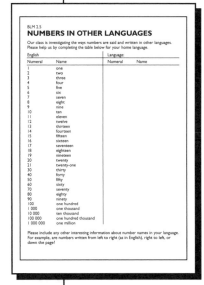

- Children can make a chart of the number words and numerals (if not Hindu-Arabic) of their first language, and use this as a reference during work.

- As a class, compare the 'teen' numbers in English with their counterparts in Chinese and other Asian languages. In English the teen numbers are said in reverse order to the way they are written as numerals – sixteen is six (units) and one ten – whereas in Chinese and other Asian languages the ten comes before the units, ten-one, ten-two and so on.

- Explore number words in other languages which show relations other than the decimal relations of our place value system. For example, in French there are vestiges of counting by 20s in the word for 80 — quatre-vingt or four-twenty.

- Explore the way numerals are written in, for example, Arabic, Chinese, and Cambodian. Hundred charts for these languages are provided on BLM 2.6. Note that Arabic is written from right to left (although each numeral follows the same place value notation as the Hindu-Arabic system), and Chinese is written vertically from top to bottom.

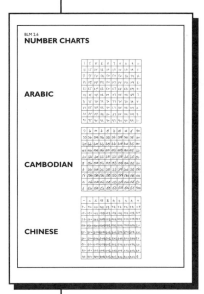

- Pose questions such as:
  - How are tens and units shown in the different languages?

- Is there a symbol for zero in each language? If so, what is it? If not, how is the difference between, say, 3 and 30 shown?

- Ask children to perform simple calculations using these numerals. They can also make up their own 'sums'.

This provides all children with the experience of working with unfamiliar symbols, and will help all children to

- appreciate the difficulties others may have in changing from their first language to English

- understand the place-value notation we use, by comparing it with other notations.

Other mathematical conventions may be different in other languages. For example:

- the decimal marker (decimal point) may be a point on the line as in 4.5, a point above the line as in 4·5, or a comma as in 4,5

- the thousands separator may be a point '.' as in 3.871, a comma ',' as in 3,871, or a space, as in 3 871. The last of these, the space, is the internationally accepted usage. However, it is not universally used, even within Australia.

Check that children from non-English speaking backgrounds understand the conventions being used in your classroom, and that they are aware of any differences between English usage and that which may be used in their homes.

It is important to check children's understanding. Beware of simply asking 'Do you understand?', as students will often answer 'Yes' out of politeness or embarrassment. Instead, be specific:

- ask questions using what, where, when, how
- ask children to tell you what you just said or did
- ask children to do tasks that will show you how much they have understood

All children should be given time to think before responding to questions. This is of particular importance if English is not the child's first language.

Visualisation and visual thinking

Visualising is a way of representing experiences and ideas. These representations are internal communication with oneself, and become communication with others when translated into drawing, writing or constructing.

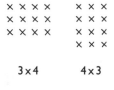

3×4   4×3

Visualisation is an important tool or process in mathematics, for the understanding of concepts and for solving problems. Many mathematical ideas (not only in geometry) can be more clearly illustrated visually than analytically. For example, by representing multiplication as an array, a visual display, it can be clearly seen by interchanging rows and columns that $a \times b = b \times a$.

Similarly, multiplying decimals or other fractions can be shown as the area of a rectangle.

3·5

2·5

$2·5 \times 3·5 = 8·75$

In measurement of length, area and volume, visualisation can help children to estimate more accurately. Children can imagine a reference measurement with which they are familiar — for example, the length of the basket ball court — and use this as a reference for estimating other lengths.

Although visualisation is a valuable skill, it is important that particular images do not become so entrenched that children regard these images as the concept itself. For example, if the numbers from one to ten are invariably presented as particular dot patterns, some children come to believe that the 'domino pattern' shows five, and the 'five dots in a line' is not five.

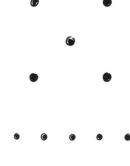

Similarly, while multiplication of decimals can be represented by the area of a rectangle, in fact most applications have nothing to do with area or with rectangles. The model is an analogy and an application, not the concept itself.

Visualisation is closely related to graphic communication, which will be examined in greater detail in chapter 6. Many activities which involve whole body movement, for example *Footsteps over the City* (page 49), also promote visualisation. Some of these activities are described in chapter 4, in the section on active communication.

The following activities will all help children to visualise mathematical problems and concepts. The first focuses on visualisation in space.

PAPER FOLDS

Years 1–7, ages 6–13
- Each child folds a piece of paper in half. Starting and ending on the fold line, they draw a shape on one half of the paper.
- Say to the children:
  - If you cut out this shape and unfold the paper, what will you see?
  - Now draw what you think you might see.
  - Now cut out your shape. Compare it with your drawing.
- Discuss with the children the process they have gone through.

Assessment
- Do children visualise the cut shape accurately?
- Do children improve as they repeat the activity with different shapes?

Adapted from Baker & Baker, 1991

The next activity deals with visualisation of number.

SEE A NUMBER

Years K–3, ages 5–9
- Say a number, e.g. 15. Ask children to draw what they see when think about that number. More than one response can be given.
- Alternatively, ask children to think of a special number, and to record the picture they see in their mind.

Assessment
- Do children use a range of images to think about the numbers?
- Do children draw on a variety of facts about the numbers? For example, how the number is made up from other numbers, or special uses for the number, such as three wheels on a tricycle.

The next two activities focus on visualisation of measurement.

BLM 2.7
ESTIMATING DISTANCE
You will need a metre ruler or a stick one metre long.

At school
Think of a distance at your home, for example, the length of the garden path, of your bedroom, of the garage.

I am thinking about the length of _____

I estimate the length to be _____

At home
Look at the distance you have been thinking of. Make another estimate of its length.

My second estimate is _____

Look at your metre stick or ruler. Now make a third estimate.

My third estimate is _____

Now measure the actual distance using your metre stick or a tape measure.

The actual length is _____

Write about the activity. What helped you to make your first estimate? Your later estimates? What would you use next time to help you estimate another distance?

ESTIMATING DISTANCE

Years 3–7, ages 8–13
- Ask children to write down their estimate the length of the [library, basketball court, passage] without going to the area.
- Now ask them to go to the area and estimate again.
- Children make a third estimate after looking at a metre rule.
- They then measure the length.
- Repeat the activity using a different distance in the school, or at home. See BLM 2.7 for a home activity.

Assessment
- Do children's estimates improve as they go through the activity?
- Do children's estimates improve when they repeat the activity?

SHAPE AND SIZE

All levels, all ages
- Ask children to imagine each of the following, and then draw (without looking at the object) the item as close as possible to its correct size:
  - a circle the size of a 20 cent coin
  - a rectangle the size of a (standard) playing card
  - a circle the size of a button on their shirt or pants
  - a square the size of the side of a die (as used in a specific classroom game)
  - a square the size of an MAB 100 (flat) block

Use other examples which will be familiar to the children, but are not visible at the time.

- Children discuss and display their work. They compare their drawings with the actual items.
- Give children a playing card to paste next to their drawing. They can make a coin rubbing to compare with their drawing.

Assessment
- Note children whose estimates were extremely inaccurate. These children will need more consolidation and practice with familiar objects.
- Repeat the activity, using the same or different items at a later date, and note whether children's estimates of size have improved.

Children need to be involved in the kinds of activities suggested in this chapter, activities which:

- extend their experience of mathematics
- extend their use of the four language modes in mathematics
- allow them to explore mathematical vocabulary
- are a rich source of information for the teacher

Such activities contribute to assessment and to building up profiles, as well as to planning further experiences which will continue to extend children's views of mathematics.

3 | SPOKEN LANGUAGE AND ACTIVE COMMUNICATION

The activities outlined in this chapter certainly dispel the myth that mathematical learning is a desk-bound activity which only involves pencil, paper and the manipulation of symbols. These activities embody the principles of engagement: doing, thinking, reflecting, organising, to name a few. They give children opportunities to explore mathematics through talk and action, and they offer alternative perspectives on and insights into mathematics. They are designed to engage children's attention and they help children to make connections between the experiences and concepts in mathematics.

The range of spoken language

The importance of spoken language, the ability to articulate and explain meaning, is viewed as a highly desirable skill which children need to develop throughout their education. Developing spoken language in mathematics is more than a preparation for formal, written or symbolic mathematics. It is an important part of concept development, and of the development of the more generalised mathematical skills — describing, explaining, justifying. Through spoken language, children learn from each other, and clarify their own thinking.

While children are engrossed in an activity, whether on their own, or playing with others, they talk. They talk in a procedural way: 'Put this in here, and then on the top put a red block.' They talk in an imaginative way: 'This building is going to be the biggest building in the whole world.' They talk in monologues, commenting on their activity, taking sometimes more than one part including noises of cars, animals or machines. When playing with others, or involved in structured activities at school, talk may move between suggestion, procedure, explanation, commentary and narrative. As they talk, children are planning and solving problems in informal ways.

Teachers can build on the talk which already occurs in the classroom, and extend and focus it to enhance learning. Organisational and teaching strategies are also necessary, to give children practice, direction and demonstration in using spoken language effectively. Talking with the whole class, talking in small groups, talking in pairs, and talking one to one with you are all important and will happen at different times during the day and week.

Spoken language encompasses more than talk within the classroom. Table 3.1 shows how the traditional uses of spoken language in the mathematics classroom can be expanded to include other audiences, exploration of mathematical ideas and problems, formal and informal reporting on work in progress or completed, drama and role play.

Table 3.1: Spoken language

	RECEPTIVE **Listening**	**EXPRESSIVE** **Speaking**
Traditional practice	to the teacher	answering closed questions
Expanded practice	to each other to parents	to each other to parents to other members of the school community discussing open-ended problems drama and role play explaining and justifying questioning others presenting reports

Audiences, once only the teacher, can expand to include other children within the class, the family, and other members of the school community. Spoken language can be used not only to answer the teacher's questions but to discuss, explain and justify, to question others, to work with children in other year levels (cross-age tutoring), and to dramatise or role play stories and situations which involve mathematics. Role play, recount, narrative and puzzles, as well as reports, are different ways oral presentations can be made. All these have value in broadening the ways children think about mathematics, and in relating the ideas and concepts of the discipline to children's own lives and interests.

At the beginning of a topic, talk is needed to establish what children already know, and what their attitudes and interests are. During the study, talk is needed to clarify, describe, plan, explain, predict and justify. Children learn from each other as they talk together about their work, and as they listen to more formal presentations from other children.

There are a number of strategies for encouraging meaningful and focused talk about mathematics. Some are more appropriate for whole class discussion, some for small groups, and some for interacting with an individual child. The first strategy described was used with a whole class. It could also be used with a small group.

Finding out: a case study – discussing fractions

We were asked to work with a year 4 class on fractions because the teacher was expressing anxiety about the children's knowledge of fractions in relation to the school's testing program.

We decided it was important to begin by finding out the children's knowledge and understandings of the topic.

The children's experiences listed in Table 3.2 clearly show that being familiar with the mathematical terminology of fractions does not necessarily mean that children understand fraction concepts or their use. Should the teacher simply present children with the terminology and mechanistic formulas in order to tackle the formal testing; or should she offer children a broad range of experiences that will put fractions into contexts — contexts that build on children's experiences and extend their knowledge, and that will provide them with the skills to use fractions in a range of situations?

A framework for facilitating and recording the discussion was developed (based on a framework for discussing literary texts, described by Aidan Chambers at a reading conference in Warrnambool in 1988). A similar structure can be used for investigating mathematical concepts, to give an idea of children's current understanding. This framework, which we have called Finding Out, provides opportunities for children to respond freely and express both mathematical understandings and concerns.

FINDING OUT

Three columns were drawn up on the blackboard, headed:

WHAT WE KNOW   WHAT PUZZLES US   PATTERNS

The children were asked to contribute initially to the first column, 'What we know'. All contributions were accepted, whether 'correct' or not. Children need to know they can contribute freely, without the fear of being 'wrong'.

Finding out afterwards whether what the children have listed in 'What we know' is indeed correct or true is an important part of this activity. In some cases, the same or similar statements may appear in more than one column – what some children know may puzzle others, and much mathematical knowledge (whether of fractions or other topics) may be perceived as 'pattern'.

After a number of statements had been listed in the first column, the children were asked to contribute to the second column, 'What puzzles us'. Finally, the children proceeded to identify patterns they were aware of in the area of fractions.

The list the children drew up is shown in Table 3.2. What is shown here does not reveal the richness of the talk and interaction between the children, and between the children and the teacher, which occurred as the list was developed.

Table 3.2: Finding out about fraction knowledge

What we know	What puzzles us	Patterns
A proper fraction is bottom heavy	Changing fractions into mixed numbers	Number lines can show fractions
Mixed number has a whole number and a fraction	Changing mixed numbers into fractions	Ordering fractions: $\frac{1}{2}$ $\frac{1}{4}$ $\frac{1}{8}$ $\frac{1}{16}$
A fraction can't be top heavy	Changing fractions into decimals	Doubling the bottom number halves the fraction
An improper fraction is top heavy	Why decimal fractions are called decimal fractions	$\frac{1}{2} = \frac{2}{4} = \frac{4}{8}$ doubled top and bottom
Fractions can't be turned upside down	All the different names for kinds of fractions	
How to change decimals into fractions	Why fractions were ever invented	
A $\frac{1}{4}$ of 60 is 15	How to write fractions	
$\frac{3}{4}$ of 60 is 45		

The list clearly shows that the children's concepts were driven by mathematical terminology. For example the children were concerned about the 'different names for kinds of fractions'. They had little understanding of the meanings and uses of fractions, for example in the statement that 'fractions can't be turned upside down'. There are also some contradictory statements, for example the two statements 'a fraction can't be top heavy' and 'an improper fraction is top heavy' both appear.

The children were certainly familiar with mathematical language but of what use was it to them in practice? The only example they gave connecting fraction knowledge with everyday experience arose from their knowledge of the divisions of the clock face.

'A $\frac{1}{4}$ of 60 is 15, and $\frac{3}{4}$ of 60 is 45.'

Further discussion confirmed that the children had little practical understanding of fractions, nor were they able to apply rules for manipulating fractions. Skemp (1976) has distinguished between instrumental understanding as 'knowing how', and relational understanding as 'knowing why'. These children had neither instrumental understanding nor relational understanding of fractions. Knowledge of mathematical terminology does not of itself mean that understandings have developed.

The list, however, provided meaningful starting points for discussion and development of fraction concepts. After rereading the list the children chose to investigate:

1. Why were fractions invented? (a purpose for mathematics)

2. Why are decimal fractions called decimal fractions? (history of mathematics)

3. Changing mixed numbers into fractions. (computation)

The three different aspects which were investigated contributed to broadening children's perceptions of mathematics. Giving the children the opportunity to choose areas for investigation in no way prevents the teacher from deciding on other questions for investigation.

Finding out: a framework for assessment

The same approach can be used effectively to draw children's learnings together after a unit of work has been completed. The framework was used with another year 4 class at the end of a unit of work which had been based on the book *My Cat Likes to Hide in Boxes*. As well as a number of other activities, the children had made boxes and investigated the surface area and the volume.

Table 3.3: Finding out: assessing learning

What I know	What I don't know/ What puzzles me	Patterns
How to measure length, width and depth	How to make a net for a box	The net for a cube is a cross
How to make a box 5 cm by 7 cm by 9 cm	Relation between a litre and a cubic metre	The area of a triangle is half the area of a rectangle
Area is measured in square centimetres		A thousand big MAB cubes fit into a cubic metre
Volume is measured in cubic centimetres		A million little MAB cubes fit into a cubic metre
100 cm = 1 metre		The area of a rectangle is length multiplied by width
How to change metres into centimetres		
You can't multiply centimetres by metres		

In contrast to the first example, these statements were driven by the children's experiences and the new knowledge they had gained through the activities. For example they knew 'how to measure length, width and depth'. They were able to make generalisations after reflecting on these experiences. After this project on making boxes, the children suggested other questions for investigation, investigations which would examine relationships, *Relation between a litre and a cubic metre*, and those which would further develop practical skills based on their experience and new knowledge, *How to make a net for a box.*

These children were able to explain their statements, using both mathematical terminology and informal language, both of which are needed and used in mathematics. For example, they drew a diagram to show why the area of a triangle is half the area of a rectangle. They were able to demonstrate their understanding of the terminology, and to move between the practical and specific example, *How to make a box 5 cm by 7 cm by 9 cm*, and the general statement, *Volume is measured in cubic centimetres.*

FINDING OUT

All levels, all ages

Use the *Finding Out* framework when you start a new topic.

• Write the topic name (for example, Multiplication, Fractions, Decimals, Graphs) at the top of the chalkboard or a large sheet of paper.

• Under this, make three columns, headed

WHAT WE KNOW   WHAT PUZZLES US   PATTERNS

- Start with the first column, asking children to brainstorm what they know about the topic.
- Move to the second column, asking children to contribute what puzzles them about the topic. Allow children also to continue contributing to the first column.
- When both these columns contain a number of statements, ask for contributions to the third column.
- Ask children if they can see connections between the statements in the list. Take coloured pens or chalk, and mark these connections.

There are a number of ways you can use this chart to promote discussion and investigation:

- You may find that the same statement has been written in two columns: for example, 'What we know' and 'What puzzles us'.
- You can ask the children who have contributed to the 'What we know column' to explain their statement to the rest of the class.
- You may also find that contradictory statements have been written. For example, in the fraction work quoted above, the two statements 'a fraction can't be top heavy' and 'an improper fraction is top heavy' appear.
  Ask the children if they can clarify the two statements, and correct or refine them so that they no longer appear to be contradictory.
- Investigate some of the statements 'What puzzles us' column. This can be done as a class, individually or in small groups.
- Ask children to clarify statements in the 'What we know' column which are unclear or appear to be misconceptions. The statement 'Fractions can't be turned upside down' is an example. Refine such statements with the children.
- Investigate the patterns children have contributed. Discuss these, ask children to explain them, and look for other patterns.

Whole class discussion

Whole class discussion can extend over a large part of a working session, as in the case studies described above, or it may be a shorter part of a session The discussion may:

- find out what the children know, and what they want to find out
- generate ideas for classroom activities or for independent projects or investigations
- bring together ideas generated by individuals or small groups
- review work in progress
- clarify tasks
- share findings
- discuss differences in methods, assumptions or solutions

One problem with whole class discussion is ensuring the participation of all children. One strategy is to ask children to write down something relevant before starting the discussion. Using the framework above, this could be 'something I know about fractions'.

In another situation it could be an estimate for the length of the corridor, or the money raised by a sausage sizzle. Each child then has something to contribute initially, and therefore becomes a stakeholder in the outcome of the activity.

DISCUSSING METHODS AND PROCESSES

One purpose of whole class discussion is to focus on methods and processes.

- Challenge all children with a short problem focusing on a concept or skill you wish them to learn.
- Ask each child for their solution.
- List these on poster paper or whiteboard.
- Then ask each child to explain their method. Discuss the differences in solution and in method.
- Do some children want to change their solution or their method as a result of the discussion?

The following problem was written on the board for a year 3 class:

$$54 - 28$$

When answers were called for, two columns of the chalkboard were filled with their solutions. All were written up, without comment. Ticks were put beside those which were given by more than one child.

'Which one is right? Or are they all correct?'

The children agreed that there could be only one right answer. A few changed their answers, after seeing the other possible solutions.

Each child in the class was asked in turn to explain how they reached their answer, and brief notes on these methods written up on the board. This is not always necessary, but in this particular case it was important that all children contributed, so that the teacher would gain an understanding of how the children were thinking.

A great variety of non-standard methods were used. Correct solutions were reached by methods which included rounding the numbers and then adjusting (e.g. $54 - 30 = 24$, add on the 2, gives 26), adding on ($28 + 2 = 30$, add on 20 to reach 50, then 4 to reach 54; 26 added altogether), using doubles ($25 + 25 = 50$ add on 4 for 54, take off 3 for the 28; 26).

Incorrect answers were reached both by faulty reasoning and through errors in addition and subtraction 'facts'. Some children applied an algorithm incorrectly, others used a non-standard method of splitting up the two numbers, but got confused about whether they should add or subtract the extra bits. Several children at this stage said things like:

'I took 4 from 8, but I shouldn't have, I should have borrowed a 10 and taken 8 from 14.'

By the time all children had contributed their method, most were convinced that the correct answer was 26. However, they were asked to work it out again, using a different method from their first attempt. They used materials such as paper and pencil, drawing, counters, squared paper, MAB blocks, and other material around the room.

By the end of the session, all the children were confident about their own methods and answers. Several correct and useful methods had been contributed by the children and demonstrated to the class, including the decomposition method (which was the preferred method according to the syllabus at the school), complementary addition (sometimes known as shopkeepers' method), and rounding up methods. These were written up and displayed for future use.

SUBTRACTION METHODS

Years 3–7, ages 8–13

You can adapt the question and the topic for other levels, or to focus on other processes.

- Ask a subtraction question, for example 54 – 28. Make this suitable for your class.
- Write up each child's answer without comment.
- Then ask the children to explain how they got their answer. Write up these methods.
- Discuss: Which answer is correct?
- When this has been established, ask the children to do the calculation again, using a different method from their first attempt. They do this whether their first answer was correct or not.
- Discuss the methods used.
  - Which methods are correct?
  - Which are easy to understand?
  - Which are easy to use?
- Each child can make up other subtraction questions, for a partner to solve, choosing from the methods presented.

Assessment

- Are children able to explain their method clearly?
- Are children able to refine their method after discussion?
- Are children able to adopt and/or adapt another method after discussion?
- Can children identify advantages and disadvantages of different methods?
- Note children who do not have an acceptable method after discussion and experimentation.

TIME FOR SHARING AND DISCUSSING

This is vital. Depending on the problem and on the group of children, this can be at the beginning, during the solving process, at the end, or all of these. This is when children learn from each other as well as clarifying and explaining their own ideas, justifying their methods and solutions, and comparing and evaluating the methods and solutions of others.

REPORTING ON THEIR WORK

Children should have opportunities to share their work, their methods and their results, during and at the conclusion of an investigation. As work is in progress, the opportunity to share what they are doing gives children a reason to reflect on and articulate their methods. By listening to others and clarifying their own thoughts, they can get confirmation that they are on the right track, or they may gain other insights or ideas which they can incorporate into their own work.

Talk in pairs and small groups

CO-OPERATIVE GROUP WORK

Small group work encourages discussion. As children work together, they use the language of hypothesis, explanation and justification. They also express their attitudes and emotions.

Children in years 6 and 8 had been working on a chessboard problem: How many grains of corn will there be on the last square of the chessboard if you start with one on the first square, double that on the second square (2), double again on the third square (4) and so on up to the 64th square.

As children worked in pairs, they formed hypotheses, made predictions, looked for shortcuts, noticed patterns, and expressed their feelings. We heard statements such as:

'I think it should be 64×64 because there are 64 squares.'

'If we multiply by 256 we'll get the square below. We can move down the side to the 64th square.'

'The numbers all end in 2, 4, 8, 6, 2, 4, 8, 6 . . . We'll never have 0 on the end.'

'I'd hate to be the poor person who had to count them — we've got over 100 million already!'

THE KING WHO WAS TIRED OF WAR

Year 5 up, age 10 up

- Show the children a chessboard and pieces. Discuss the game. What is the object of the game? How do the pieces move? If possible, arrange time for children to play the game, and to teach each other the rules.
- Tell or read the story of the invention of chess (see BLM 3.1).
- Stop the story when the king asks his treasurer to calculate the amount of corn needed for the inventor's reward.
- Ask the children to estimate, and then work out, the number of grains of corn on the last square of the chessboard, as a preliminary to finding the total.

The children can work in pairs on the problem.

- When the children have reached a solution, ask them to share their answers and their methods. Ask questions such as:
- Did you try any shortcuts? Did the shortcuts work?
- Did you find any patterns?
- Was the calculator useful?
- Read the end of the story. Compare children's answers with the answer in the story.
- Children can try reading the numbers (eighteen quintillion, four hundred and forty-six quadrillion, seven hundred and forty-four trillion, seventy-three billion, seven hundred and nine million, five hundred and fifty-one thousand, six hundred and fifteen — using the US convention that one billion is one thousand million).
- Children can write about the activity, reflecting on what they have learnt.
- You can extend the activity by asking children to find out the volume or mass of the corn.

Assessment

- Did children make hypotheses or look for shortcuts?
- Did children check their hypotheses or short methods?
- Did children find patterns as they worked?
- Did children persist to find a solution?
- Note children's comments, both spoken and written. What additional information do they give about the children's understanding?

BLM 3.1
THE KING WHO WAS TIRED OF WAR

Once there was a king who was tired of war. He longed for another occupation, which would be just as interesting and challenging, but which would not mean death and destruction.

He therefore offered a reward to the person who could find him an occupation as absorbing and challenging as leading his armies into battle.

A scholar named Sessa invented the game of chess and brought it to the king. The king was delighted with the game, its complexity and subtlety.

He asked Sessa to choose any reward.

'Thank you, your Majesty,' said Sessa. 'All I wish is this: as many grains of corn as it would take to fill the sixty four squares of this chessboard, putting one grain on the first square, two on the next, then four, eight and so on, doubling each time.'

'Surely,' said the king, 'the game is worth a better reward than that! Think again.'

'No,' said Sessa, 'I am a modest man, and that will satisfy me.'

So the king called his vizier to calculate the amount of corn needed, and to bring the bag of corn to Sessa.

Away went the vizier, to set his reckoners to work on the calculation.

STOP HERE. HOW MANY GRAINS OF CORN DO YOU THINK WERE NEEDED ON THE LAST SQUARE? HOW MANY ALTOGETHER?

It was hours later that the vizier appeared again before the king.

'Well,' said the king, 'have you the bag of corn which Sessa has requested for his reward?'

'Your Majesty,' said the vizier, 'it is not possible. All the grain in your kingdom will not be enough. Even all the granaries of the world do not hold the amount you have promised.'

'How can that be? How many grains are needed?'

'Your Majesty, the last square of the chessboard would need 9,223,372,036,854,775,808 grains of corn, and the total number of grains on the board would be 18,446,744,073,709,551,615.

'What's to be done?' asked the king. 'Since I cannot pay this reward, what do you advise me to do?'

'There is one way,' replied the vizier. 'Tell Sessa he can have his reward, if he counts the grains of corn himself. In his whole lifetime, working day and night, he would be unable to count but a small fraction of the number of grains before he died.'

Much has been written on the benefits of group work (see for example, Baker & Baker 1986; Dalton 1985). There is no doubt that there is more opportunity for individuals to talk in small groups than in a whole class situation, but it is important to ensure that the talk is useful and focused. Children need support and practice to work effectively in groups, as well as needing tasks which they find interesting and motivating. Children who are not clear about what is expected of them, who are not motivated, or whose progress is not monitored, may be off task for much of the working time.

Gooding and Stacey (1993) found that effective groups were those where there was more talk, and that this talk included explicit discussion of mathematical concepts and conventions, reading out loud, repeating each others' statements, proposing ideas, giving explanations, refocusing discussion, and responding to questions.

In another recent research study (Cooper et al. 1993), children who were assigned roles for small group problem-solving tasks performed significantly better than children who were not assigned roles.

To ensure that all members of the group contribute, and that the group maintains its focus, you can use a number of strategies. For example:

- Make sure the task is clear, and that children know what is expected of them.

- Assign roles: a recorder, a reporter, a gopher (who 'goes for' any equipment needed), a checker (who checks that the group remains on task) and so on. You need to introduce these roles gradually so that children understand the responsibilities associated with each.

- Children write down their initial ideas before they come together as a group. The first task of the group is to listen to everyone's ideas. This ensures that all children contribute to the group.

- Tell the children that you may call on any member of the group to report on their solution. This means that all children have to be sure they understand what they have done. They may need time to practise their report in the group before presenting it to the whole class.

- If groupings are not working because of dominant children, change the groupings to put the dominant children together. Other mismatches can be altered, for example you may need to adjust the ability levels within some groups.

- Spend time establishing good working groups. Give the groups the opportunity to monitor their own working through self-assessment. They can comment on the role of each group member as well as commenting on such things as:
  - What we did well as a group
  - What we need to do better
  - One thing we agreed we learnt about mathematics

- Provide small group activities which focus on active listening skills. See, for example, the activity *Make it like mine* on page 45.

- Provide positive feedback to groups which are working well. Be specific: 'Everyone in this group is listening well today', or 'Writing your ideas on a large sheet of paper was a good idea. It helped you to see what to do next'.

CROSS-AGE TUTORING

A buddy system, where older children work individually with younger children, will give both children opportunities to talk about mathematics. The need to explain clearly to the young child will help the older partner to clarify and articulate his or her own thinking. The younger partner will benefit from the one-to-one contact and from the different perspective offered.

PRESENTING REPORTS

On completion of an investigation or project, children can present their reports within their own class, to another class, or to a whole school assembly.

DRAMA AND ROLE PLAY

There are many opportunities for drama activities in mathematics lessons. The experience of being *within* the mathematical situation, rather than looking at it from outside is a powerful force in the learning process. Drama and role play will usually involve both spoken language and physical activity. Physical involvement is discussed below (pages 46–52).

Drama can be a small part, or the major part, of a session or sequence of sessions. Children can act out number rhymes or songs, such as 'There Were Ten in the Bed', or a story such as 'The Wolf and the Seven Little Kids' (p. 17), before solving a problem, playing a game or recording the mathematics arising from the rhyme or story. Older children might present a dramatic version of 'The King Who Was Tired of War' (p. 42) after working on the problem of the number of grains of corn on the chessboard. Other examples of rhymes, poems and stories which involve drama activities as part of the mathematics learning can be found in Griffiths & Clyne (1988, 1990, 1994).

Children may present their solution to a problem as a role play or drama. For example, children can dramatise the way they would spend a given amount of money. In the MCTP lesson 'Tell Me a Story' (Lovitt & Clarke 1988) students are asked to interpret a graph showing the level of the bath water and act out the story behind the graph. Look for other graphical representations which 'tell a story'.

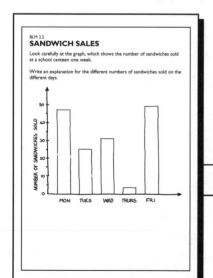

BLM 3.2
SANDWICH SALES
Look carefully at the graph, which shows the number of sandwiches sold at a school canteen one week.

Write an explanation for the different numbers of sandwiches sold on the different days.

SANDWICH SALES

Years 1–7, ages 6–13
• Present the graph on BLM 3.2, which shows the sandwich sales at a school canteen over a week.
• Ask the children to look carefully at the graph, and explain why the sales of sandwiches change as they do.

Assessment
- Do the children read the information on the graph correctly (e.g. the number of sandwiches sold each day)?
- Can the children identify possible reasons for the changes in sales from day to day?

LISTENING GAMES

Games are an excellent way of developing spoken language skills. Some games, such as the one below, focus particularly on precise descriptive language, and active listening.

MAKE IT LIKE MINE

Years K–7, ages 5–13

In this activity, two children are separated by a screen. Each child has identical materials. These can be counters, pattern blocks, construction materials, 'junk' items, geoboards, or pencil and paper, etc.

- One child builds a model, draws a picture, or makes a pattern (as appropriate for the material used).
- S/he then describes what s/he has done, so that the second child can do the same.
- In an alternative version, the second child asks questions as s/he tries to replicate the model.

Assessment
- Are the children able to communicate so that the models match?
- Do the children use unambiguous language as they communicate?
- Note use of everyday and mathematical language.

Although the above activities focus on the development of spoken language, other language modes are also present. Providing opportunities for children to discuss, observe, challenge, justify and explain what they are doing is helping them to clarify meaning and at the same time increase their understanding of mathematics. Including activities like these in the mathematics classroom also helps you to find out more about children's understandings.

Home involvement

Parents and other caregivers and family members can be involved in talking about mathematics with their children. This may be reading a book and carrying out a mathematical investigation or talking about the mathematics in the book. It can be playing games which explore or use mathematics. It can be doing 'the problem of the week' sent home for parents and children to solve together. Family maths nights at the school, where parents and their children enjoy maths activities together, help to develop positive attitudes towards mathematics for both parents and children, as well as developing positive relations between the school and the home.

Before sending home activities to be done as a family, it is important to meet with parents, or write to them, so that they are aware of the importance of their involvement.

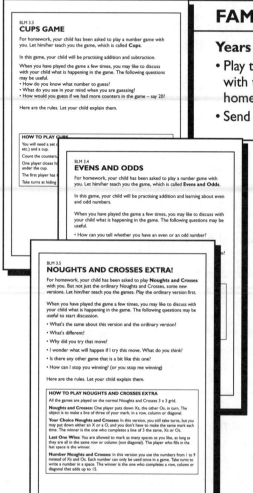

FAMILY GAMES

Years K–7, ages 5–13

• Play the game chosen in class, so that children are familiar with the rules (included on BLMs 3.3–3.5) before taking them home. However, do not discuss the strategies at this stage.
• Send a game home for 'homework'

Active communication

Active communication in the receptive mode involves *interpreting* the actions of others. These actions may be demonstrations of

• how to use measuring instruments
• manipulation of concrete materials
• whole-body activities such as drama and mime, or pacing the circumference of a circle to find the relationship with the diameter

In this context, the activities may be modelled by the teacher, and the students replicate the action, or a group of students may be involved in the action, while others observe.

In the expressive mode, it involves *doing* all the above: manipulating materials, measuring, demonstrating, performing, and participating in whole-body activities. It also involves deciding *how* to do these things.

Table 3.4: Active communication

	RECEPTIVE Interpreting	EXPRESSIVE Performing and demonstrating
Traditional practice	interpreting the teacher's demonstrations	manipulating materials measuring
Expanded practice	watching and interpreting the actions of other students	demonstrating drama role play mime physical activity

The importance of using a variety of concrete materials to enable a range of representations of mathematical concepts and problems to be made is discussed in chapter 2. The value of practical work in measurement and in space is well recognised. Drama is also discussed above. This section focuses particularly on whole-body, physical involvement.

Discovering by walking around a large circle, as well as by drawing and measuring circles on paper, that the circumference of a circle is three-and-a-bit times the diameter adds an extra dimension to students' understanding of the ratio π. The physical activity is likely to be remembered longer, particularly if the activity includes discussion and reflection.

Physical involvement adds kinaesthetic learning to the cognitive, providing an extra dimension to mathematics learning. The opportunity to explore the mathematics from inside, to be a part of the mathematics, creates a personal involvement. Lovitt & Clarke (1988) suggest that 'a very strong visual image remains with the participants, and ... this image can be tapped in follow-up work inside the classroom.'

Physical movement provides opportunities to express attitudes and emotions. Body language is often an unconscious expression of attitudes and feelings. Activities such as drama, role play and mime offer opportunities to express feelings consciously. Children enjoy activities and games which involve whole-body movement, so it is likely that their interest and participation will be high.

Every mathematical topic offers opportunities for some physical involvement. Usually, an activity undertaken on a large scale outside can be repeated on a small scale indoors, through drawing or manipulating materials, thus providing another perspective and consolidating the learning in a more formal way.

Fine motor activities such as finger counting and body counting, and computing using finger methods, provide alternative perspectives to pencil and paper methods and the manipulation of concrete materials. Sign language, used by the deaf, employs movement as well as position to indicate numbers. Investigating this language would be of interest to all children, and of particular benefit if you have deaf children in the class or in the school.

The following suggestions for activities span the range of mathematical topics and ages.

NUMBER CARDS

Years K–3, ages 5–9
- Make a set of number cards, 0 to 10, which can be hung around the children's necks. There are many activities which you can do with these cards. While ten (or eleven) children wearing the cards are doing an activity, the other children can be involved, observing and making suggestions. For some activities, you may wish to have the whole class wearing numbers.

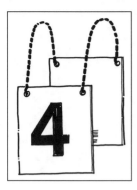

- Children put themselves in order.
- Children sort themselves into even and odd numbers.
- Children pretend they are house numbers or letter boxes, and put themselves in 'street' order.
- In twos or threes, children make the biggest (or smallest) 2- or 3-digit number they can.
- Children make sums: find pairs which add to make a given number, for example pairs which make 9.
- Children make up new activities for the number cards.

Assessment
- Note children who have difficulty with particular activities.
- Are children able to devise their own activities using the number cards?

Adapted from the MCTP activity 'Maths in Motion' (Lovitt & Clarke 1988)

MATHS WALK

Years 2–4, ages 7–10
- Draw a grid 5 x 5 on the asphalt, similar to the one shown below, and large enough for children to stand in. The object is to move from START to FINISH, trying to reach a given total, for example 30. The players must move to an adjacent square (including diagonals).
- Children form teams of 3 or 4. One team member stands on the START. The other team members direct him/her to move.
- Change the total to be aimed for.
- Children can make up other games to play on the grid.
- After the game outside, they can make similar grids on paper, to play in the classroom.

4	7	6	6	5
START	8	5	8	2
2	4	1	0	6
8	1	9	3	FINISH
4	0	3	7	9

Assessment
- Do children co-operate in their team, or does one child dominate?
- Do children think ahead, or do they just consider each move in isolation?

MAKING POLYGONS

Years 1–7, ages 6–13
You will need between 10 and 20 metres of string for each group and lots of space.

- Discuss the names and properties of different polygons. Explain the meaning of the word *polygon*, which is derived from the Greek *poly-* meaning many, and *-gon* meaning angle. The usual definition is a many-sided figure, with the assumption that all sides are straight. The first polygon is the triangle.
- Ask the children to use the string to make different polygons. For example, to make a triangle, three children hold the string taut. You can then ask them to make a regular or equilateral triangle by moving until they are equal distances apart; a right-angled triangle by making one corner 'square'.

They can make a range of quadrilaterals, for example: square, rectangle, rhombus, parallelogram.

- Upper primary children can use a large protractor to measure the angles made, for example 90°, 60°, 45°, and the interior angles of pentagons, hexagons, etc.
- Allow time for discussion of the activity, and follow up by children drawing the polygons on paper.
- There is the opportunity here to make a Topic Glossary (p. 23), or a Book of Polygons.
- Children can write about the activity and what they have learnt from it.

Assessment

- Do children have intuitive understanding of the shapes they are making?
- Do their drawings indicate that their understanding has improved after the activity?
- What do children think they have learnt?

Adapted from the MCTP activity 'Regular Polygons' (Lovitt & Clarke 1988)

FOOTSTEPS OVER THE CITY

Years 3–7, ages 7–13
You will need a scale map of your city or local area, preferably one which can be displayed to the whole class.

- With the children, list some important city landmarks (between five and ten), choosing ones which are spaced across the city and suburbs.
- Ask children to locate the landmarks on the map, and to work out their direction and distance from the school.
- Allocate each landmark to a different child, who will need to check the distance and direction again. One child is allocated the school as a landmark.

- The class then goes outside, the student who is representing the school is placed first, and the direction of north established. The other landmarks then place themselves in the correct relative positions, using a scale of (say) one footstep to one kilometre.
- The rest of the class then have to guess which landmark each child represents.
- The activity can be repeated with more and/or different landmarks, and with the 'school' placed in a new spot in the playground. Make sure all children have a turn as 'landmark' and as observer.

Assessment

- Were children able to use scale and direction when locating landmarks on the map?
- Could children place themselves correctly on the outdoor 'map'?
- Could the children who were observing identify the different landmarks children were representing?

Adapted from the MCTP activity 'Footsteps across Australia'
(Lovitt & Clarke 1988)

DRAWING A CIRCLE

Years K–4, ages 5–10

- Children 'draw' a circle in pairs. One child stands still holding one end of a piece of string (at the centre) while the other child walks around the centre keeping the string taut.
- Ask the children watching to describe what is happening, and to identify the path made by the walker.

Assessment

- Do the children identify the path as a circle?
- Do the children realise that the walker (the outside edge of the circle) is always the same distance from the child inside (the centre)?

CIRCUMFERENCE OF A CIRCLE

Years 3–7, ages 8–13

- Children draw a circle, as described in the activity *Drawing a circle* above.
- Children can then work in pairs, making circles of different sizes. They can investigate the relationship between the radius (the length of the string) and the circumference (the distance walked). Before measuring, children should make an estimate.
- Discuss the process and the results.
  - Did every group get the same results?
  - How could the ratio be found more accurately?
  - What difference will it make to use bigger or smaller circles?
  - What would be more accurate measurement techniques (for example, tape measure, or trundle wheel rather than pacing)?

- Follow up by drawing and measuring circles on paper, and measuring the radius and circumference of circles found in everyday objects such as cans, jars and wheels.

Assessment

- Do all groups find a reasonable approximation for the ratio of circumference to radius? (e.g. a bit more than 6, or $6\frac{1}{3}$, or between 6.1 and 6.4)
- Can children identify factors which would help them to get a more accurate result? (For example, using bigger circles, using a tape measure.)

A MATHEMATICAL MIME

Year 5–adult, age 10 up

- Ask children to mime a mathematical concept, for example a fraction, π, an acute angle, or a quadrilateral.

  Children can work individually, or in small groups.

This is an extremely challenging task. The discussion which follows will be a major part of the learning, as children explain their actions in terms of their intentions and the concepts they were trying to convey.

Assessment

- Did children focus on surface aspects of the concept (e.g., the shape of π), or did they attempt to show the meanings of the concepts (e.g., that fractions are made up of equal parts).
- Did children co-operate to present their mime?

A MATHS TRAIL

Year K–adult, age 5 up

- Plan a maths trail around the school grounds or in the school neighbourhood. Make sure that there are a range of concepts and skills involved, not only number, and that questions can be answered using a variety of methods.

  This can be done in a number of ways.

  - A family maths trail is an excellent way of involving the school community. Provide copies of the trail, with space for answers, comments and methods. Ask families to return their completed copy by a certain date, and award certificates to participants. Provide feedback to the school community by publishing comments and answers which are unusual or interesting in the school newsletter.
  - An older class can plan a maths trail for younger children.
- Provide some examples of the kinds of questions that might be in a maths trail. The children can work in pairs or small groups, with each

group making up three or four interesting questions. As a class, you can then pick the final questions for the trail, and work out a route. Make sure that every group has at least one question in the trail.

• Each group refines their question(s), to ensure that they are clear.

• The older children can partner the younger children on the trail.

Assessment

• Could children make up interesting questions for the trail?

• Was there a variety of questions to choose from?

• Did participants enjoy the maths trail? Were they challenged by it?

• Did older children partner younger children successfully, for example could they provide sufficient support but ensure that the younger children engaged with the mathematical thinking needed?

Other measurement activities with physical involvement are included in chapter 6 and in *Maths Makes Sense* (Griffiths & Clyne 1994). The MCTP Activity Bank (Lovitt & Clarke 1988) has a chapter on physical involvement in mathematics learning.

4 | EXTENDING THE MODES OF WRITTEN LANGUAGE

If children are to write meaningfully and effectively about mathematics, they must have quality material to write about and quality experiences which have purpose and which will support them in their writing tasks. This means that children need:

- activities which engage and challenge them
- activities which focus on mathematical concepts and skills
- explicit discussion of mathematical techniques and features
- models for writing, both fiction and nonfiction
- explicit discussion of writing techniques and features

By providing the range of reading and writing experiences suggested in this chapter, together with interesting, relevant and challenging mathematical experiences, you will enable children to write about mathematics, and extend and enrich their mathematical understanding as they do so.

Reading and writing in mathematics and mathematics in reading and writing

We are constantly reminded by those within and by those outside the education community of the importance of students being able to read and write. Reading and writing are essential skills for learning mathematics in our culture, and children need to be given opportunities to read and write about mathematics in ways that will extend their understandings of mathematics.

It is generally accepted that we need language to learn and to use mathematical knowledge. However, there is not enough attention given to the fact that we also use mathematical understanding as we read.

As you read the newspaper article 'Lead levels worry doctors' on BLM 4.1, list the mathematical knowledge you need to understand the content and implications.

The 218 words of text in the article are replete with cultural knowledge, including mathematical concepts and terminology. We are not suggesting that all competent readers have a deep understanding of every concept. However, some understanding of all of them is necessary to make sense of this article. Some of these concepts are so well understood and automatically used by most readers that they would not even recognise them as mathematical.

We have identified the following mathematical ideas in the article. Compare what you have found with our list.

- numbers and numerals, including decimals
- proportion (as in two to five times)
- rates (micrograms per decilitre, grams a litre)
- average
- dates (in the 1970s, by mid-1994)
- vocabulary such as more, most, reducing, mid-, above, down to, concentration
- measures (gram, microgram, litre, decilitre)

BLM 4.1
LEAD LEVELS WORRY DOCTORS

Melbourne: Children in Australian cities have two to five times the blood lead levels of American children because of more lead in our petrol, a doctor said yesterday.

Community pediatrician Dr Garth Alperstein said most children living in Australian cities had lead levels between 10 and 20 micrograms per decilitre of blood.

"On a population basis that's concerning — we can do a lot better," he said.

"In the US, preschoolers in the 1970s had an average level of 15, but that's now down to four, mainly as a result of reducing lead in petrol."

Above 10 is the new United States' level of concern.

Dr Alperstein works for the Central Sydney Community Health Services and Royal Alexandra Hospital for Children. He and six other medical experts said in *The Medical Journal of Australia* that plans were urgently needed to do something about environmental lead.

The doctors found blood lead levels of children living in an area near a lead smelter were almost identical to those of children in other areas whose main exposure was lead in petrol and paint.

The doctors said it would be worthwhile for the petroleum industry to make a public commitment to reduce lead concentration in leaded petrol by mid-1994 to that of most European countries — 0.15 grams a litre.

The *Courier-Mail*, Brisbane, Monday 5 April 1993

LEAD LEVELS WORRY DOCTORS

ratio, proportion

Melbourne: Children in Australian cities have two to five times the blood lead levels of *time* American children because of more lead in our petrol, a doctor said yesterday.

vocabulary

vocabulary of comparison

Community pediatrician Dr Garth Alperstein said most children living in Australian cities had lead levels between 10 and 20 micrograms per decilitre of blood.

rate

"On a population basis that's concerning — we can do a lot better," he said.

vocabulary — statisti

date, time

"In the US, preschoolers in the 1970s had an average level of 15 but that's now down to four, mainly as a result of reducing lead in petrol."

number

vocabulary — comparing

Above 10 is the new United States' level of concern.

Dr Alperstein works for the Central Sydney Community Health Services and Royal Alexandra Hospital for Children. He and six other medical experts said in *The Medical Journal of Australia* that plans were urgently needed to do something about environmental lead.

number

equality

The doctors found blood lead levels of children living in an area near a lead smelter were almost identical to those of children in other areas whose main exposure was lead in petrol and paint.

vocabulary

rate

The doctors said it would be worthwhile for the petroleum industry to make a public commitment to reduce lead concentration in leaded petrol by mid-1994 to that of most European countries — 0.15 grams a litre.

time, date

The *Courier-Mail*, Brisbane, Monday 5 April 1993

LINKING MATHEMATICS AND READING

For parent or teacher groups: use the article on lead levels (BLM 4.1), or find a suitable one of your own choice.

For year 5 up, age 10 up: use the article on rabbits (BLM 4.3), or find a suitable one of your own choice.

- Provide copies of the chosen article.
- Ask participants to underline any words and phrases in the article which require mathematical knowledge for understanding.
- As a group, share and classify the words and phrases marked.

Parents and teachers can discuss where these concepts might fit in the primary and secondary school curriculum.

They may also like to examine the article on lead levels for science and technology knowledge.

- Children can use the article as a starting point for investigation into one or more of the topics listed by the group, or into the problem of population growth. See *The rabbit problem*, page 63.

BLM 4.1
LEAD LEVELS WORRY DOCTORS

Melbourne: Children in Australian cities have two to five times the blood lead levels of American children because of more lead in our petrol, a doctor said yesterday.

Community pediatrician Dr Garth Alperstein said most children living in Australian cities had lead levels between 10 and 20 micrograms per decilitre of blood.

"On a population basis that's concerning — we can do a lot better," he said.

"In the US, preschoolers in the 1970s had an average level of 15, but that's now down to four, mainly as a result of reducing lead in petrol."

Above 10 is the new United States' level of concern.

[...] ealth Services
[...] er medical
[...] re urgently

[...] area near a
[...] ther areas

[...] n industry to
[...] leaded petrol
[...] rams a litre.

BLM 4.3
RABBITS!
A lifestyle that's geared to growth

As the breeding season approaches, the female rabbit (doe) joins a small group of up to three males (bucks) and up to seven other does in a well-defined territory. Once pregnant, she carries the foetuses for about 30 days.

In the last week of pregnancy she digs a special breeding hollow, or stop, in which to give birth. This is usually a domed chamber at the end of a tunnel. If her status is high in the group, the stop will be deep in the warren. There her kittens will be safer from predators than those of does lower on the social scale who have to dig stops where they can.

A few hours before the birth, the doe works feverishly to line the stop with dry grass. An hour before the birth she plucks fur from her belly and thighs and uses it to cover her naked, newborn kittens. The birth takes five minutes, after which she emerges from the stop and seals the entrance with soil. She may now immediately mate and become pregnant again.

The doe visits her kittens once or twice a night for no more than five minutes at a time. But because rabbit's milk is richer than both cow's or goat's milk, this is enough to provide them with nutrition. At three weeks the kittens emerge to eat grass but may still approach their mother to suckle occasionally.

Parental concern doesn't last long. If the doe is pregnant, she must prepare for her new litter and, within a few days, she drives her kittens away.

The wild rabbit's breeding capacity is legendary. In harsh environments, like northern Europe, this counters the high death rate — sometimes as high as 90 per cent of the young — caused by predators, climate and disease.

Rabbits breed in winter and spring when green feed is available. There is a strict hierarchy among both sexes in the breeding groups, status being won in fights and displays of aggression. The dominant bucks and does protect the group's territory, which they mark out with scent.

The dominant does, about 25 per cent of the females, produce more than half the young. The second and biggest group (about 43 per cent) produces all but seven per cent of the rest.

The doe can become pregnant at three months. Litters vary from three to seven kittens and she may have as many as nine litters per season — though five or six is the average. In normal conditions she may produce 25 young in a season. Before the season ends, does from her earlier litters may also be producing.

Rabbits, which generally live for two to three years, can go without drinking as long as their food contains at least 55 per cent water. Green vegetation usually has between 70 and 80 per cent. Dry plants have between five and 25 per cent and when there is nothing else, rabbits die.

Rabbits stop plants from regenerating by eating their seedlings. In parts of Australia the result is that various plant species, particularly the woodland trees and tall shrubs, are destined for extinction. In parts of north-east South Australia, even in areas where livestock is absent, cassia, mulga and other acacias are seriously threatened and have died out in places.

In linking mathematics and reading, we are seeing a two-way process. Reading contributes to learning mathematics, and mathematics contributes to understanding what we read. Only if we understand the mathematics will we be able to move into making predictions or discussing implications for the future — processes which are utilitarian as well as creative and imaginative. If we integrate mathematics and reading in this way, we will be setting up contexts for problem solving which integrate social, moral and mathematical thinking.

NEWSPAPER MATHS

Years 4–7, ages 9–13

Have a daily paper available for one or two weeks.

- Each day, share the newspaper among the children, and ask them to identify any mathematics.
- Children cut out articles, highlight and write about the mathematics they find, and make a collage mural of Newspaper Maths.
- Small groups or individuals may like to follow up some mathematical issues which arise from this activity.

Assessment

· Do children identify a range of mathematics concepts in the newspaper, or do they identify only number examples?

> • Are children able to describe the different kinds of mathematics they encounter in the paper?
> • Do children identify problems and investigations which arise out of the articles they have collected?

Table 4.1 shows that reading is much more than simply reading a textbook, or reading exercises or problems from the blackboard, and writing is more than sums and algorithms. In this chapter, we explore some of the ways in which reading and writing broaden and deepen children's understanding of mathematics.

Table 4.1: Written language

	RECEPTIVE Reading	EXPRESSIVE Writing
Traditional practice	sums word problems	'fill the gap' 5 + 7 = algorithms
Expanded practice	narratives factual texts interactive texts procedural texts newspapers and magazines environmental print each other's writing mathematical texts	descriptions recounts explanations reflections stories journals factual creative and imaginative

The communication modes are often used together. When working, children move freely between speaking, listening, reading, writing, constructing and drawing. So, as reading is discussed, you will be able to identify many applications which also involve talking, writing, graphics and active involvement.

Reading

Just as stories and non-fiction texts provide contexts for exploring personal issues and social issues, investigating language, developing understandings about our environment and so on, they also provide contexts for exploring mathematics and for solving problems using mathematics. Texts can:

• provide the initial stimulus for an investigation

• illustrate or broaden a unit of work

• bring together the learnings at the end of a unit

Using a range of texts contributes to and complements the practical work, games, everyday problems, mathematical investigations and written and oral work that are part of good classroom practice.

Children need a wide variety of texts and text styles which support and extend their learning in the area of mathematics, as they do in other curriculum areas. Although there are relatively few books for children which focus specifically on mathematics, many books use mathematics to provide information or to tell a story. If children encounter texts that present mathematics in a variety of contexts, and if at the same time activities and explorations are carried out which make the mathematics explicit and meaningful, children's understanding of mathematics will be extended and enriched.

Books, and other written material such as posters, magazines and newspapers, are needed to:
• stimulate children's interests
• present new perspectives
• help children to find answers to their questions
• help them to pose new questions
• give them models for communicating information to others

Different kinds of written material can play different roles in developing mathematical thinking.

NARRATIVES

The main role of narrative in the learning of mathematics is in relating mathematics to the human world, and in particular to the child's world. Stories may illustrate mathematics concepts, pose problems, suggest investigations, or provide models for children's own writing. They demonstrate the relevance of mathematics to the child and to society.

THE CROW AND THE PITCHER

Years 2–7, ages 7–13
• Read or tell Aesop's fable.

A thirsty crow found a pitcher with some water in it, but she could not reach the water with her beak. She thought she would die of thirst, within sight of water!

But she devised a plan.

She collected some small stones, and dropped them, one by one, into the pitcher.

Slowly the water level rose, until it reached the top of the pitcher.

She was able to reach the water with her beak, and satisfy her thirst.

• Provide a tall, thin, glass or clear plastic measuring cylinder or vase. Half fill the container with water.

- Pose the problem:
  How many stones do you think we will need to raise the water to the top?
- Children write down their estimates. There will probably be some discussion on how big the stones should be.
- Now collect stones from outside. You may like to ask each child to collect a handful, or send a group to collect a container full.
- Ask the children to look at the stones that have been collected. Do any of them wish to change their estimates?
- Now carefully place stones in the container, making sure that children record the number that go in.
- Allow opportunities for children to revise their estimates if they wish. They should keep a record of all their estimates, not erase their earlier estimates.
- When the water reaches the top, record the total number of stones.
- Ask the children to compare with their estimates. Whose original estimate was closest? Ask the children how they made their original estimates.
- In small groups, children repeat the activity, using different containers and/or different levels of water.
- Ask children to write and draw about what happened as they did the activity.

Assessment
- Do children make reasonable estimates?
- Can children justify their estimates?
- When working in small groups, can children record and keep track of the numbers efficiently?
- Do children's estimates improve when they repeat the activity?
- Can children explain what has happened clearly?

The next activity, on magic squares, is introduced with a Chinese legend. Introducing children to mathematics developed in different regions of the world and by different cultures is an important way of investigating the origins of mathematical thought.

MAGIC SQUARES

Years 3–7, ages 8–13
- Read or tell the legendary origin of the magic square (BLM 4.2).
- Show the children the pattern on the magic tortoise, and ask them to interpret the pattern.

Explain that the numbers all have special meanings. The even numbers are the feminine or yin numbers, and the odd numbers masculine or yang. The five in the centre represents the earth, and the four elements surround the earth: 4 and 9 represent metal, 2 and 7 fire, 1 and 6 water, and 3 and 8 wood.
- They can then undertake the tasks and investigations given on BLM 4.2.

Make sure that all children understand that the columns, rows and diagonals of a magic square add to the same total before children start making their own magic squares.

Extensions

Magic squares are found in many books of mathematical puzzles and investigations, for example in Lorraine Mottershead's *Metamorphosis* (1977). Kit Williams' *Masquerade* (1979) uses magic squares as clues to the treasure hunt which is the basis of the story.

Assessment

• Are children able to identify number patterns in the lo-shu?

• Are children able to make their own magic squares?

• Could children explain how they made their magic squares?

• Note different methods children used to make their magic squares.

• Could children explain why a 2 x 2 magic square is impossible?

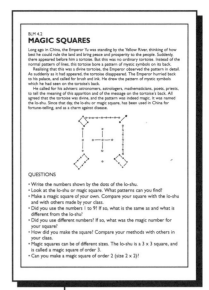

FACTUAL TEXTS

Topics popular with children such as Dinosaurs and Space are rich in mathematics. Factual books on topics such as these can provide both starting points and the mathematical information needed for projects and investigations. Texts such as *The Book of Animal Records* and *What Did You Eat Today?* (Drew, 1987, 1988) will stand alone as resources for mathematical investigations, or, for more extended work, can be supplemented by other factual texts. (See Clyne & Griffiths, 1990, 1991). *The Guinness Book of Records* is a wonderful source for mathematical information and investigations, and a volume which captures the imagination of children.

Children should also have access to a range of reference material as they investigate mathematical problems. For example, the year 3 children who were investigating the cost of feeding all the cats in Melbourne for one day used a street directory and an encyclopedia to find specific information.

VETS IN AUSTRALIA

Years 3–7, ages 8–13

• Pose the problem:

How many vets are there in Australia? (or your state or city)

• The children work in small groups on the problem. Have available resources such as street directories, atlas, year book and encyclopedia.

• When children complete their solutions, each group has to justify their solution. The children can vote on the most convincing solution.

Assessment

· Do groups identify relevant information needed?

· Are children able to find the information needed, or make good estimates, within the group itself, by making a survey, or by using reference books?

· Can children perform the necessary calculations to produce an answer?

· Are children able to justify their solution?

INTERACTIVE TEXTS

Interactive texts demand active response from the reader. These texts may be in the form of games or puzzles, or may involve prediction or experiment. Some are narratives, for example *The Eleventh Hour*, and *Cumberland Road*. The pictures in *Anno's Magical ABC* are to be viewed in a curved mirror, and the book provides instructions for making your own anamorphic drawings.

Puzzle books abound. Kirsch & Korn's *Number Games* has a range of interesting puzzles which will be enjoyed by individuals and small groups, or can be the basis of class activities. Marion Walter's *Magic Mirror* challenges children to create symmetrical pictures by using a mirror. *Creature Features* will involve young children in counting and thinking about shape when guessing the creature in the box. *Alone in the Desert* and *The Cat on the Chimney* focus on solving problems using technology and, in the process, use many mathematical concepts and skills.

PROCEDURAL TEXTS

Procedural and instructional texts also involve children in applying knowledge. These texts link practical activities to conceptual knowledge. They also link the knowledge to social purposes, for example, making a bird feeder, baking a cake, constructing a musical instrument, playing a game, or writing secret codes. Teachers use procedural texts in many cases to focus on the particular language patterns associated with these texts. However, it is important that the children are also involved in making or doing, in carrying out the instructions, so that the purpose of the text is realised.

RULE-A-CODE

Years 2–7, ages 7–13
- Write up the instructions for this secret code on a poster (or make your own blackline master).
  - *To write secret messages in Rule-a-code you will need a ruler marked in centimetres.*
  - *Place the ruler on your paper, and put a mark at the zero of the ruler.*
  - *Now write your message, putting one letter at each centimetre mark of the ruler.*
  - *Add extra letters between those in the secret message*
  - *To read the secret message, you will need a ruler marked in centimetres.*
  - *Place the ruler on your paper, with the zero at the mark on the left.*
  - *Now write down the letter at each centimetre mark.*
  - *Read your message.*
- After making the code and sending messages, children can identify and describe the mathematics they used to create messages and to decode the messages.

Assessment
- Do children use the ruler correctly and accurately?
- Can children identify and talk about the mathematics they have used?

NEWSPAPERS AND MAGAZINES

Numbers, graphs, percentages, tables of figures, abound in newspaper and magazine articles. Use these as illustrations of the uses of mathematics, examine them critically for the all too frequent errors in or misuses of mathematics, and make these the start of problems and investigations.

A year 6 class read about the breeding cycle of rabbits in *Geo* magazine (see BLM 4.3), and investigated the problem:

How many rabbits will there be at the end of a year, starting with one pair of rabbits?

These children started from the information given in the article, rather than from Fibonacci's simplified version. They needed to decide for themselves what parts of the article were relevant, what assumptions they needed to make, and how they were going to represent and solve the problem.

For example, they had to decide whether to take the maximum number, seven, the minimum, three, or an average number of kittens per litter, and whether to allow for the maximum number of litters, nine, per season, or the average, five or six.

This is a well known problem, first discussed in the 12th century by the Italian mathematician Leonardo Fibonacci (also called Leonardo of Pisa). It is usually presented in the simplified form that Leonardo used:

A certain man put a pair of rabbits in a place surrounded on all sides by a wall. How many pairs of rabbits can be produced from that pair in a year if it is supposed that every month each pair begets a new pair which from the second month on becomes productive?

The number of rabbits at the end of each month (assuming none die) follows this sequence, called the Fibonacci sequence:

1   1   2   3   5   8   13   21   34   55   89   144

They had to decide whether to assume equal numbers of does and bucks in each litter. This was important when the does in each litter reached breeding age.

They discussed which months of the year were likely to be the non-breeding months, and decided that this was likely to be in summer, when the vegetation in most parts of Australia dries out. This contrasts with the situation in Europe, where less food would be available in winter.

They used a variety of representations to solve the problem. This is a problem which has no single correct answer; there are many variables involved. The process of determining what these variables are, making reasonable assumptions about them, and systematically following through the implications of those decisions, is the process of solving 'real' problems.

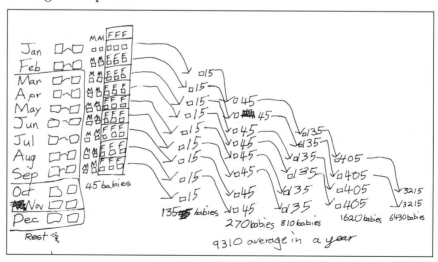

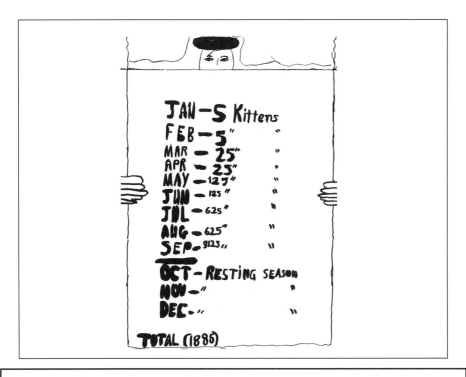

THE RABBIT PROBLEM

Year 5 up, age 10 up
- Have children read the article on rabbits (BLM 4.3).
- Discuss the information contained in the article, with particular reference to the mathematical concepts and terminology.
- Pose the problem:
  How many rabbits would one pair of rabbits produce in twelve months?
- Children can work in pairs or small groups on the problem. They will need time to discuss the problem, and to decide what assumptions to make about, for example, the breeding season, the number of litters, the number of kittens in a litter, the proportion of males and females, whether the rabbits breed in the first month and so on.
- Ask the children to write a report on their solution, and to present this to the class.

Assessment
- Can children identify the relevant information from the article?
- Do children consider the range of factors involved, or do they only consider one or two factors?
- Do the children make reasonable assumptions and simplifications to enable them to solve the problem?
- Do children work systematically to solve the problem?
- Can children explain their working and their solution clearly?

This article can also be used for children to identify, describe and classify the mathematics needed to understand it. See *Linking mathematics and reading*, page 55.

BLM 4.3
RABBITS!

A lifestyle that's geared to growth

As the breeding season approaches, the female rabbit (doe) joins a small group of up to three males (bucks) and up to seven other does in a well-defined territory. Once pregnant, she carries the foetuses for about 30 days.

In the last week of pregnancy she digs a special breeding hollow, or stop, in which to give birth. This is usually a domed chamber at the end of a tunnel. If her status is high in the group, the stop will be deep in the warren. There her kittens will be safer from predators than those of does lower on the social scale who have to dig stops where they can.

A few hours before the birth, the doe works feverishly to line the stop with dry grass. An hour before the birth she plucks fur from her belly and thighs and uses it to cover her naked, newborn kittens. The birth takes five minutes, after which she emerges from the stop and seals the entrance with soil. She may now immediately mate and become pregnant again.

The doe visits her kittens once or twice a night for no more than five minutes at a time. But because rabbit's milk is richer than both cow's or goat's milk, this is enough to provide them with nutrition. At three weeks the kittens emerge to eat grass but may still approach their mother to suckle occasionally.

Parental concern doesn't last long. If the doe is pregnant, she must prepare for her new litter and, within a few days, she drives her kittens away.

The wild rabbit's breeding capacity is legendary. In harsh environments, like northern Europe, this counters the high death rate — sometimes as high as 90 per cent of the young — caused by predators, climate and disease. Rabbits breed in winter and spring when green feed is available. There is a strict hierarchy among both sexes in the breeding groups, status being won in fights and displays of aggression. The dominant bucks and does protect the group's territory, which they mark out with scent.

The dominant does, about 25 per cent of the females, produce more than half the young. The second and biggest group (about 43 per cent) produces all but seven per cent of the rest.

The doe can become pregnant at three months. Litters vary from three to seven kittens and she may have as many as nine litters per season — though five or six is the average. In normal conditions she may produce 25 young in a season. Before the season ends, does from her earlier litters may also be producing.

Rabbits, which generally live for two to three years, can go without drinking as long as their food contains at least 55 per cent water. Green vegetation usually has between 70 and 80 per cent water. Dry plants have between five and 25 per cent and when there is nothing else, rabbits die.

Rabbits stop plants from regenerating by eating their seedlings. In parts of Australia the result is that various plant species, particularly the woodland trees and tall shrubs, are destined for extinction. In parts of north-east South Australia, even in areas where livestock is absent, cassia, mulga and other acacias are seriously threatened and have died out in places.

ENVIRONMENTAL PRINT

Junk mail catalogues, food and drink cartons and tins, timetables and signs, can all be read and used in the mathematics classroom. As this year 4 child wrote:

'Two weeks ago we had to bring a supermarket magazine and we had to write a suitable breakfast, lunch and tea, and work out the cost of it, Then we had to write a shopping list and the prices out of the magazine.'

JUNK MAIL CATALOGUES

All levels and ages

Junk mail catalogues are a wonderful resource for mathematical activities.

- Making menus

  Collect supermarket catalogues. Ask children to plan their meals for one day, using only items found in the catalogue. They then write their shopping list, and the cost of the day's food.

- Choosing presents

  Collect toy and department store catalogues.

  Ask children to work out a shopping list for presents for every member of their family. They will need also to work out the cost. You may decide to put a limit on the amount to be spent, or you may leave it open.

- Sorting and classifying

  In small groups, children choose and cut out an item from each page of a catalogue.

  They then sort and classify the items, and paste them into their groups. Each group of children reports on the way they have classified the items.

- Making up problems

  Ask the children, working in small groups, to make up their own problems using a catalogue. They then give their catalogue and their problems to another group to solve.

Assessment

- Can children read prices accurately?
- Can they calculate totals and change correctly?
- Are children able to sort items into logical classifications? Can they justify the way they sort items?
- Are children able to make up interesting and challenging problems?
- Note the mathematics children use in the problems they make up.

CHILDREN'S WRITING

When children have completed a task, give them time to share what they have done with others. It is not possible for every child to read their writing aloud at every session, but other strategies can be used.

- Put children in groups of 4 to 6. After they have read each piece of writing aloud within the group, one child's writing is chosen to be read to the class.
- Put children's writing on display, so that over a period of time others can read the work.
- Make class books which can be borrowed in school or taken home to share with families.

MATHEMATICAL TEXTS

The mathematics register in which conventional mathematics texts are written uses a very different syntax from everyday English writing. Children need specific knowledge and skills to be able to effectively read and understand mathematical texts.

Mathematics is written very concisely. Every word and symbol is important to the meaning. This contrasts with many other texts, where many words can be omitted, and the reader can still make general sense of the text. This means that children must learn to read accurately, often with pencil and paper to check their understanding. For example, $7 - 4 + 2$ has a different meaning from $7 - (4 + 2)$.

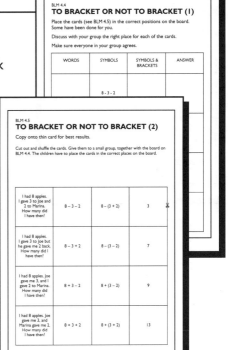

TO BRACKET OR NOT TO BRACKET

Years 3–7, ages 8–13
- Pose the following problems

  I had seven dollars. I bought a toy car for \$4 and a notebook for \$2. How much money was left? Write an equation to show how much money I had left.

  I had seven dollars. I bought a toy car for \$4 and then my nana gave me another \$2. How much money was left? Write an equation to show how much money I had left.
- Children share and compare the different equations they have written. If some have used brackets, highlight their use. If children have not used brackets, ask them to rewrite one of their equations so that it is still conveying the same message, but uses brackets.
- Make up 'stories' to match each of these equations:

  $10 - 5 - 2 = 3$

  $10 - (5 - 2) = 7$
- Children share and compare the different 'stories' they have written.
- Working in pairs, each child writes an equation using brackets. The partner has to make up a story for the equation.
- Match equations and stories. Provide each small group with a copy of the base board (BLM 4.4) and a set of the sixteen cards from BLM 4.5. One statement in each row has already been filled in.

The task is to place the remaining cards correctly on the base board.

While everyday English is read from left to right, mathematics is not always read in that way. For example, 6)12 is read as 'twelve divided by six', not 'six divided by 12'. The numeral 16 is read as sixteen, not as 'ten six' or 'onety-six'. Some of the most commonly used algorithms for addition, subtraction and multiplication are worked from right to left. While some children pick up without difficulty the different order in which mathematical text may be read, for many children this is a source of confusion, which needs to be dealt with explicitly. Reading mathematical text aloud, and explaining what the text means, are important.

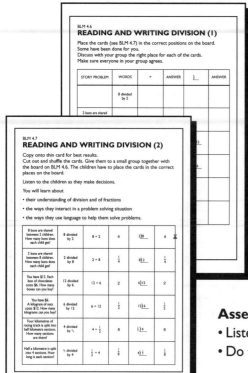

READING AND WRITING DIVISION

Years 4–7, ages 9–13

• Provide each small group with a copy of the base board (BLM 4.6) and the cards from BLM 4.7. One statement in each row has already been filled in.

The task is to place the remaining cards correctly.

This task is designed to confront children with common misconceptions about division, for example what the conventional symbols mean, whether the order matters, and how to divide a smaller number by a larger number. Through group discussion, these misconceptions can be addressed.

• When the group has completed the task, ask for a report from the group. Then ask each child in the group to write about what they learnt in doing the task.

Assessment

• Listen to the groups as they discuss the task.
• Do they explicitly discuss the order of the operation, fractional answers, division of a smaller number by a larger number?
• Do they listen to each other, and give all group members opportunities to put their point of view?
• Do they give explanations?
• Do they ensure that the discussion is focused on the task?

Adapted from Bell (1986) as described in Gooding & Stacey (1993, p.42).
See p.43, for a summary of Gooding & Stacey's findings on small group discussion.

Writing

Writing helps children to clarify and extend their understandings. In writing about what they are doing, and what they have found out, children continue to organise information and then reflect on their knowledge and understanding. Writing:

- provides records for the teacher, which assist in the assessment of the child's learning and understanding
- provides a focus for discussion with the child
- helps in planning for future curriculum activities.

Writing also expresses attitudes and emotions. Personal writing, for example in journals or letters, and creative writing, as in narrative or poetry, have their place in the curriculum, both in broadening children's concepts and understandings, and in relating the learning to the children's interests and experiences.

Writing activities cover a wide range of forms, from informal listing and notes to formal reports and displays for presentation to a wider audience such as parents and the community. Writing may include text features such as graphs, diagrams, tables, equations, maps and drawings. These are discussed more fully in chapter 5.

The texts which children have been reading, independently or with the teacher, provide models for children's own writing. Discussion and comparison of different genres, styles and text features will make explicit the writing features which children may incorporate for different effects or different audiences. Teachers need to model different kinds of writing, such as narrative, report, explanations, procedure, and different formats for presentation of information, such as graphs, diagrams, flow charts, or maps.

Children can write at the beginning, during and at the end of mathematical activities. For example, they can list, individually, in small groups or as a class, what they know and what they would like to find out about a topic at the start of a unit of work.

During the work, they may write informal notes as they make observations, perform calculations, or represent mathematical problems in a range of ways, they may add to their lists of things they know, or clarify and redraft the questions they have posed. They may write a daily or weekly journal in which they respond freely to the learning, or they may compile more structured responses to activities. They may compile tables, draw and label diagrams, make graphs, draw maps.

At the end of the topic under study, children may write reports, posters, stories, or a class book. These final presentations are likely to incorporate features of the writing they have been doing earlier.

As they write in these different ways, children clarify their thinking, organise their knowledge, identify questions to investigate, communicate with others, and provide a record for themselves and others of the work they have been involved in.

Writing about mathematics, and using mathematics in original and creative writing, have gained much attention recently. Writing can take many forms. Leah Richards (1990) found the following types of writing in her year 7 mathematics class in the space of one week:

summaries	labels	descriptions
translations	instructions	predictions
definitions	notes	arguments
reports	lists	explanations.
personal writing	evaluations	

MONITORING WRITING IN MATHEMATICS

BLM 4.8
MONITORING WRITING IN MATHEMATICS

Checklist	MON	TUES	WED	THUR	FRI
Summaries					
Translations					
Definitions					
Reports					
Personal writing					
Labels					
Instructions					
Notes					
Lists					
Evaluations					
Descriptions					
Predictions					
Arguments					
Explanations					

Over a week, record the different kinds of writing children use in mathematics. Note down any comments and plans.

All levels and ages

This is an interesting activity to undertake in your classroom. If you are able to do this in conjunction with another teacher, you can discuss your findings and plan together.

By monitoring what children write, extending their writing experiences, and presenting your findings to the children, you will help children to become more aware of a broader range of written forms, and not see mathematics simply as sums and algorithms.

• Use BLM 4.8 to record the different kinds of writing children use in mathematics.

• Share what you find out with other teachers as well as with the children.

Assessment

• Use what you find out to plan mathematical activities to include other kinds of writing.

• Note particular children who may need support in extending the range or quality of their writing.

WRITING ABOUT A MATHEMATICS ACTIVITY

One teacher, who was committed to a problem-solving approach, found it difficult to monitor each child's or each group's progress. By asking the children to record how they had solved a problem, she was able to keep track of their progress, and also used the writing as a basis for discussion later if time ran out during the session. She found that the writing became more than a record of what they had done. It became a vehicle for them to clarify their understanding and reflect on their learning.

When asked to write about a mathematics activity, children's responses will show a great variation. After listening to a poem about measurement, children chose materials to measure the height of a chair seat, a book, and other items in the classroom. They then wrote about the activity.

'55 counters measure a seat of a chair.'

'Jeremy and I found out that 13 Unifix was the same as *Cyrus the Unsinkable Sea Serpent*.'

'We read a book with Mrs Griffiths and it was fun.'

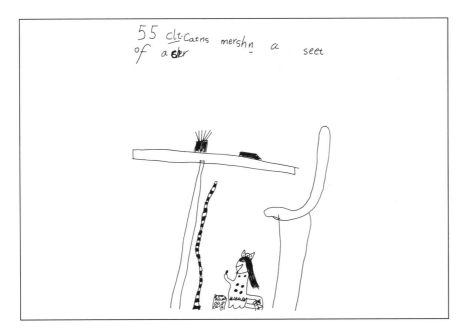

All three year 1 children have written simple recounts. However, the children who produced the first two pieces of writing focused, as was the teacher's intention, on the measurement activity. The child who wrote the third sentence has told us nothing about the mathematics. When young children are first asked to write about a maths lesson, their text in many cases may focus on their emotional response to the lesson. Sharing the children's writing is sometimes enough to encourage initially reluctant writers to have a go, or to focus those who have not addressed the mathematics.

If a child's writing continues to focus on peripheral information or emotional responses only, you may need to be more specific about the task. However, a range of teaching strategies is needed, from providing a clear focus to freely asking children to respond to the task.

The year 3 boy whose work appears below was asked to write about the mathematics the class had done during the year — a broad request. He chose to write about a doubling problem. His writing is a recount, which focuses more on the events than on the mathematics. However, what does come through, is a sense of wonder, enthusiasm and achievement. This is the more remarkable as it comes from a child who in the previous year had seemed uninterested, in either writing or mathematics.

> ### The Magic Doubling
> We did plant work on a doubling plant. It doubled every day. Our teacher said she was not expecting (us) to get this far. David asked how big would it get in one year. Our teacher said you can find that out. He only got as far as 4 weeks but Ryan got his computer and did the sum and he got the answer. They had 5 trips to the moon and 5 back.

Children who have had experience in writing about mathematics will gradually become more focused in the content of their writing, if offered support and direction. These year 3 children working in a co-operative group to find the number of poles in an 18-metre fence show their reasoning clearly. They have added explanation to their recount.

'First we divided 18 by 3s because there had to be 3 metres before and after a pole. But when we checked it by making a model it was a different answer because we forgot we need a pole on the end of the fence too.'

A piece of writing such as this could be presented to the class as an example of a clear description and explanation. Both the mathematics and the language can be highlighted and discussed.

The year 3 children whose work follows were solving a problem based on the book *One Hungry Spider*: how many creatures were there in the story? These recounts have a strong focus on explanation.

'I got a ten out of the MAB then nine ones. Then I got out 8 ones and I realised I needed to exchange for ten and so on until I got to 44. Then I had a bit of trouble but I managed to fix it up.

Answer 53. Equipment I used MAB.'

'First we discussed whose plan we would use. We chose Luke's. What we did was draw the number like put 1 for one spider then 234 for the birds which would = up to 4 creatures. But my [idea] was drawing the numbers at the bottom and counting across till I got to the top but we used Luke's way.'

After the chessboard activity (p. 42), the children (year 6) were asked to write about what they had done. These are no longer simple recounts. These children are beginning to reflect on what they have done.

'I thought it was interesting because I learned a bit about chess. We also learned that the calculator only went to a certain number and then it stopped. We know that because an 'E' showed up on the screen.'

'We discovered if you multiplied the end number by itself [you] got the number that goes in the row below, then you multiply the answer by the top end one till you get the bottom number.'

WRITING AS PART OF THE PROBLEM-SOLVING PROCESS

As well as writing when a maths activity has been completed, children write as they are involved in the problem-solving process. Such writing may consist of calculations, guesses, keeping a record of what they are doing, or other notes, or it may be more structured.

The children whose work follows had read *Counting on Frank*, and were checking the remarkable calculations made by the narrator.

Peas

What we know!

We know that each year has 365 days except for leap years which have 366 days.

We also know that he knocked off 15 peas every night for eight years.

What we need to know or find out.

Lauren's table 1 metre tall

Farrah's table 83 cm tall

Lauren's dining room 4m width, 6m length

Farrah's dining room 4m width, 8m length

Average table height 80 cm

Average dining room length 7m, width 4m

Our plan.

We planned out that we needed to use X and +.

We built a cubic metre and measured how many peas could fit in there.

How we worked out the result

$$\begin{array}{r} 365 \\ \times 8 \\ \hline 2920 \end{array} \qquad \begin{array}{r} 15 \\ \times 2 \\ \hline 30 \end{array} \qquad \begin{array}{r} 2920 \\ \times 15 \\ \hline 43830 \end{array} \qquad 12{,}000{,}000$$

Two girls were investigating whether 15 peas dropped on the floor every night for eight years would reach the table top. They used a framework we had suggested to give some structure to the activity:

- What we know
- What we need to know or find out
- Our plan
- Our working
- Results

Giving a structure provides some children with a starting point and direction, which they may need if they have not previously worked in this way. Other children may choose not to use the structure.

This piece of writing shows a number of features of the children's thinking, and raises a number of other questions. The children have identified key information needed to solve the problem (for example, how many peas were dropped, how many peas in a cubic metre, how high is the table, and how big is the dining room). Their use of the word 'average' indicates a limited understanding of the concept, or possibly that they have used other data which they have not recorded. A recording such as this provides valuable information for discussion with the group, and for planning further experiences.

We have also done this activity on several occasions with adults. It is interesting to note that in many cases, and with other problems, adults tend to worry about the lack of information. For example we hear statements like 'But we don't know the size of the room, or the height of the table', or questions like 'What if the peas squashed down?', 'Were they frozen peas?'. Children tend to simply accept the problem, and proceed to make estimates or measurements to supplement the information in the story. We could reflect on the difference by asking whether the adult responses may be due to the fact that they are conditioned to expect all the information to be stated in the problem.

Other children in this class did not use the framework. This boy was checking the statement that the average ballpoint pen draws a line 24 and a half metres long before the ink runs out.

'I tested the question by drawing 240 lines 10 cm long. The estimate he gave us was hopeless.'

The three boys in the next group lacked confidence and were very reluctant to get started. Eventually they decided to tackle one of the simpler problems from the book.

'The boy is about 8 years old because if he is sixteen metres tall and he grows 2 metres every year he must be 8 years old. So we think that it is right.'

With the confidence they had gained in this problem, they then tackled a more complex one: would ten humpback whales fit in the boy's house? Note the language they use: 'We think that', 'the width would probably be', 'because'. They have read for specific information: how long is a whale?

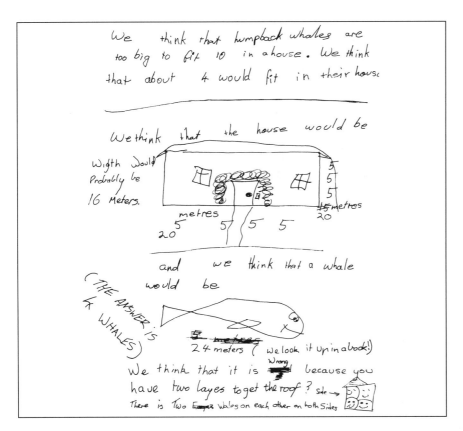

We think that humpback whales are too big to fit 10 in a house. We think that about 4 would fit in their house

We think that the house would be

Width would Probably be 16 Meters.

metres 20 5 5 5 5

5 metres 20

and we think that a whale would be

(THE ANSWER IS 4 WHALES)

24 meters (we look it up in a book!)

We think that it is Wrong because you have two layes to get the roof? side → There is Two Whales on each other on both sides

Children in year 1 (five- and six-year-olds) were given five pieces of string, all the same length. They were asked to make different shapes with their string, and to paste these shapes onto paper. They were then asked to write about the length of the different pieces of string.

The writing shows different stages of understanding about conservation of length. Some children are content to describe the shapes as different in length (although they had checked that they were the same length initially), while others are grappling with articulating the differences between their visual perception and their knowledge of the initial state.

'The smallest shape is a circle and the longest is a straight line.'

'My shapes are all the same size but they don't look like it. The longest piece of all is the straight piece. The smallest piece of all is the ball.'

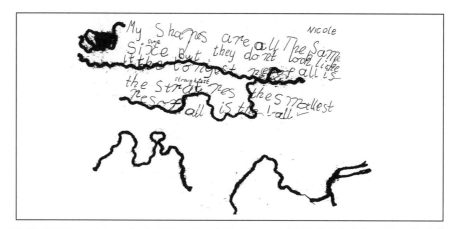

'The biggest one is the straight one. The second biggest one is the hook one. The third biggest is the two one. The fourth biggest one is the loop one. The fifth one is the triangle one. They're not all different. They're the same.'

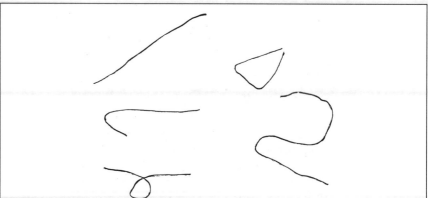

STRING ACTIVITY

Years K–2, ages 5–8
- Cut five pieces of string, all exactly the same length (between 20 and 30 cm), for each child.
- Ask the children to check that all their pieces are the same length.
- Ask the children to experiment with the shapes they can make with the string. Talk about some of the shapes they make. Ask children to describe the different shapes.
- Now ask children to make a different shape with each of their pieces of string, and to paste these onto a sheet of paper.
- Children can now write about their pieces of string, focusing on the shape and the length.

'Write about the pieces of string you have pasted down. Tell about their shape and how long they are.'

Assessment
- Do children show an awareness that the pieces of string are, or should be, the same length?
- Are children able to articulate their understanding of conservation of length?
- Are children able to describe the shapes they have made?

OTHER FORMS OF WRITING

Journals   Writing a maths journal regularly gives children the opportunity to move from a mere recount of the maths activity to a specific mathematical focus – explaining, reflecting, and perhaps hypothesising and investigating. You may wish to provide focus questions such as:

- What did you learn this week?
- What would you like to know more about?
- What was difficult for you? Why?
- In your journal today I would like you to write about…
- Can you explain the…in your writing?

Children responded in a variety of ways:

> 'Today we found out about numbers. We had to do a sheet with numbers up to 108. We learnt what prime numbers are. It was good fun. The prime numbers are so far this 1, 2, 3, 5, 7, 11, 13, 17, 19, 23, 29 ... and square numbers are 1, 4, 9, 16, 25, 36.'

> 'We are having a bit of trouble. We need one more square but we can't find one. It is very hard make a quilt. We have got twenty-five squares on our quilt. There are twelve yellow squares and thirteen purple squares.'

Imaginative writing   Writing using mathematical ideas and concepts need not be factual. Stories provide models for imaginative writing with strong mathematical content. For example, we have used *10 Apples up on Top* with young children, *Alexander Who Used To Be Rich Last Sunday* and *10 for Dinner* in middle primary, and *Anno's Mysterious Multiplying Jar* with older children. The work children have produced clearly demonstrates that mathematical and imaginative thinking is integrated, and indeed may be inseparable.

Children also use mathematics spontaneously in their writing. For example, a year 3 boy, asked to write about a recent school camp, produced a counting book based on his experiences at camp.

Integrating the curriculum in this way means that you need to be aware of what tasks children are doing. When involving children in imaginative writing as part of mathematics, you need to ensure that the mathematics has the prime focus. During the process of creating an imaginative text, encourage children to focus on the mathematics and to discuss the mathematics of the text. When this part of the activity is finished you may then choose to focus on refining the writing for presentation or publication.

If children are writing a story modelled on *Alexander Who Used To Be Rich Last Sunday*, they must ensure that the money spent adds up to $2.00 (or whatever amount is decided on). While acknowledgement should be given to children who have imaginative ideas about how they might spend or lose their money, this must not be at the expense of accurate calculation. Similarly, children should not feel inhibited by the need to produce perfect spelling, handwriting and grammar. Using mathematics in a context is the primary focus.

If a decision is made to publish or display the stories in some way, editing and proof reading should follow as a matter of course.

How Kristy spent 2.00

Today I Wanted to Do Some Chores for Mum
because I needed Money. So I asked Mum.
Mum said ok and She gave me Chores
and this is What She got me to do.
Wash the Dishes, and dry, SWeep the floor,
and tidy My room. good Work Son Mom
Said. so she gave me TWo Dollars Wow
I said TWo Whole Dollars. thats What I Wanted
So the Next day I spent it on a Book $35
and on a rubber $40 and a pet frog $20 and a pencil
$90 and a pencil case $15.

The children who wrote texts based on *10 for Dinner* exchanged their stories with other groups. The other group had to draw a graph based on the information in the story.

WELL DONE TINA & PAUL

The guests wore their best clothes to the wedding 8 wore dinner suits and 6 wore evening dresses 1 wore a tuxedo

But 1 wore a wet-suit and flippers with a snorkle.

The final example is modelled on *Anno's Mysterious Multiplying Jar.* Writing texts of this kind involves:

- following a mathematical pattern
- calculating with large numbers
- logical thinking, because each item has to follow logically from the previous item.

Try it for yourself.

From 'I went fishing', based on *Anno's Mysterious Multiplying Jar.*

Writing good innovative texts needs strong demonstration and modelling of both the writing and mathematical skills. It is important to discuss the features of the original text, and the mathematical concepts included in it. After reading *Anno's Mysterious Multiplying Jar,* which builds up a factorial pattern $(1, 1\times2, 1\times2\times3,$ up to $1\times2\times3\times4\times5\times6\times7\times8\times9\times10)$ in a magical story, children explored and discussed the number pattern. They identified the differences between a doubling pattern and the factorial pattern of the book, which some children confused initially. The concept of a billion was discussed, and the different meanings in the UK and the USA explained. Before writing individual stories, they wrote a class text together with the teacher, so that they became aware of the logical demands of the task, and some possible solutions.

Posing problems as well as solving problems involves children in thinking about and applying their knowledge. After they have solved a problem, ask children to make up another one which may be similar, or which extends the situation. These children from a year 3/4 class made up the following problem based on the last page of *The Doorbell Rang.*

'There are 14 people. Each person got $7\frac{1}{2}$ cookies. Grandma baked 65. How many did the stranger bring?'

This is a complex problem, involving multiplication of fractions, followed by a subtraction. Making up their own problem gave these children the opportunity to extend their mathematical thinking.

In posing their own problems, children need to reflect on the original problem and clarify their understanding. They need to be confident about the mathematics they are using. The problems they write will indicate their understanding of the original problem or situation, and are valuable for assessment purposes. Posing problems also allows for creative thinking and application.

Explanations to an outsider   These could be letters to a friend explaining a mathematical idea or process, or an explanation to share at home. By making the audience someone who has not participated in the class experience, the child is encouraged to make explicit the ideas which may be considered common knowledge within the classroom.

Instructions   Instructional texts and explanations can overlap. However, the main focus of an explanation is to show why something works, whereas an instruction tells how to do it. This means that the concern is with making the process clear rather than with understanding.

Posters Children can write reports of their investigations for display in the classroom, in the school, and in the local library or shopping centre. This is an excellent way to promote your school within the community. Posters should combine graphic and written communication.

Reflective writing   Ask children to reflect on work they have done over a period of time. They will need the opportunity to talk as a group or as a whole class about the work, in order to consolidate ideas. The work which follows was written by a child from a year 3/4 class.

> ### About our trees
> When we started learning about trees and measuring them I was wondering how we would do it. It was then I realised most of our work in some ways were maths. I liked finding a tree of our own and writing about it. One way we did it was looking through our legs and walk like that until you could see the whole tree. Then we measured how far it was and that would be how high your tree was. Now I enjoy maths because I know that it's not all hard sums.

BUILDING ON CHILDREN'S WRITING

Writing is not always an end point; it can be the springboard for both language and mathematical work. The following piece was written by a child in the same class as the writer of the previous piece.

> ## Working with trees
> To measure a tree, instead of climbing up it with a tape measure, you can get a cardboard piece and make another square in the middle of it, and move back until you can get the whole tree in the square. Then when you have the whole tree in the picture you make a mark from where you are standing and then measure from the tree to where you are standing and then the answer on the tape measure is how tall the tree is.

This is a clear and succinct instructional text, which can be built on for two purposes. You can use it with other students:

• to demonstrate, examine and refine the writing of an instructional text

• to articulate more clearly the mathematics needed so that the instructions can be carried out by another person (for example, the size of the square in the cardboard)

The example below outlines a procedure which can be used.

WRITING INSTRUCTIONS

Years 3–7, ages 8–13
• Do a practical maths activity such as *Drawing a circle* (p. 50) or *Maths walk* (p. 48).
• Ask the children to write instructions after they have completed the activity.
• Choose some interesting examples from those the children have written, and follow the procedure below.
  · Write the text on a poster or on overhead. Ask the children to identify the different steps in the process. Show how, by writing each step on a separate line, the instructions become clearer and are easy to follow.
  · If each step is to be on a new line, do the sentences need to be broken up? Are the steps in the right order?
  · Ask the writer and the children whether there is enough information in the text to enable another person to do the activity.
  · Is there other information which would help us to carry out the procedure? For example, a heading to provide the purpose of the activity, a list of materials needed.

- After going through one example in detail, look at some other examples more briefly to highlight other points, to present options, or to consolidate points already discussed.
- In small groups, the children rewrite their instructions, adding any additional information needed.
- The children then write instructions for another measurement activity which they have recently been involved in.

Assessment

· Can children identify the steps and the sequence necessary for the procedure?
· Do children include all essential information?
· Are children's instructions clear and unambiguous?
· You can note improvements or problems when children write instructions for another activity.

Follow-up

'Volume of a room' from the *MCTP Activity Bank* would be an excellent follow-up activity. In this, children are provided with metre rulers or sticks, and asked to estimate the volume of the room they are in. They then plan in small groups a way of finding the actual volume, using the metre sticks or other measuring equipment. After doing and discussing the measurement, children can write instructions, using their own method.

5 | GRAPHIC COMMUNICATION AND VISUAL REPRESENTATION

The ability to interpret graphical information such as tables, graphs and maps accurately is of great importance in understanding the information which is presented to us daily in books and newspapers, on television and on computer screens. Learning to communicate through reading and making graphics is essential. Equally important is the skill of critical reading of graphics. The activities in this chapter focus on the interpretation and creation of graphics. They also include discussion of the features which clarify, obscure or distort information. In this way, children are provided with a sound basis for developing critical literacy in the area of graphic communication.

Drawings, diagrams, graphs and maps are important in communicating and learning mathematics. As with other language modes, the use of graphics can be informal or formal. Often, the graphics are combined with written language, for example in tables, graphs and maps.

Table 5.1 shows how traditional practice can be expanded to include a wider range of sources and activities.

Table 5.1: Graphic communication

	RECEPTIVE Reading & interpreting	EXPRESSIVE Creating & drawing
Traditional practice	tables, graphs and diagrams presented by teacher or textbook	tables, graphs and diagrams under direction

	RECEPTIVE Reading & interpreting	EXPRESSIVE Creating & drawing
Expanded practice	each other's work	recording in own ways
	tables, graphs and diagrams from books, magazines, newspapers	representing with concrete materials
	maps	drawing
		maps
		collecting and organising data
		creating tables, graphs and diagrams from raw data

Drawing

Children's drawings can provide you with important information about their understandings of particular concepts or topics. See how these year 1 children (six-year-olds) have drawn calculators. They range from Mark's, which shows no numbers, through Stuart's, where the numbers go from 1 to 6, and are placed at random, Michelle's, where the numbers are in order — but the reverse from the actual calculator layout — to Marcus's, which shows the numbers in the correct layout, includes 0 and the decimal point, as well as all the function buttons.

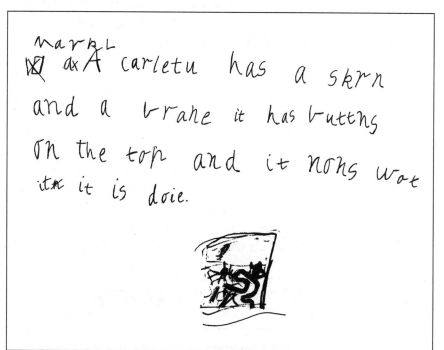

Mark

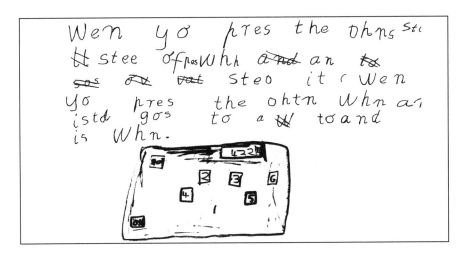

Stuart

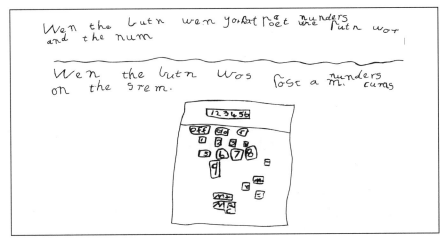

Michelle

Marcus

As children make drawings, they also learn about the things they are drawing. The year 2 child whose drawings of a clock are shown on the next page had great difficulty with his first attempt, tried again, and on his third attempt had integrated the essential features of the clock face. These three drawings were made in the same session.

EXPLORING CALCULATORS

Years K–3, ages 5–9
- Provide a calculator for each pair of children. Explain that they are to take turns with the calculator, and they must make sure that they can both see the display at all times.
- Ask the children to find out as much as they can about the calculator. They can press any of the buttons.
- Ask each pair to tell the class something they have found out.
- Now ask the children to draw a picture of the calculator. When they have finished their drawing, they can write about what they have found out.

Assessment
- Do children include all the numbers in their drawing?
- Do children include the other buttons in their drawings?
- Do children order the numbers and function buttons as in the actual calculator layout?
- Note the kinds of activities children have been trying out, and the information about calculators they report, orally and in their writing. Use these for further experiences and investigations.

This activity can be adapted to other topics, for example children can draw a clock, or a set of balance scales.

Drawing may be the most efficient way to solve a problem, to check a solution, or to use in conjunction with analytic methods. Recall the children who used division to solve the problem of the number of fence posts (p. 70), but checked their solution with a diagram, and immediately found their error. By encouraging diagrams as well as analytic representation, visual and analytical thinking are both used. This enables children to:

- make use of their preferred learning strategies
- develop and consolidate their use of alternative strategies
- match appropriate strategies to problems
- check solutions

DIAGRAMS TO SOLVE PROBLEMS

Years 2–7, ages 7–13
- Present one or more of these problems to the children.
- Ask them to solve each problem in two different ways. One method must use a diagram or picture, and the other should not.
  - *Fence post problem:*
    You have to build a garden fence 18 metres long. You need 3 metres between posts. How many fence posts will you need?
  - *Handshake problem:*
    There are 8 people at a party. If everyone shakes hands with everyone else, how many handshakes will there be altogether?
  - *Present problem:*
    Five friends go to a new year party. If everyone brings a present for everyone else, how many presents will there be altogether?
  - *Lily pond problem:*
    A gardener planted a tropical lily early one morning. She found that it doubled its size every day. At the end of the fourth day it covered half the pond. How much did it cover when she planted it?
  - *Legs and tails problem:*
    How many horses are there in the show ring if there are 12 more legs than there are tails?
- Children compare their solutions and strategies in pairs or small groups.
- Discuss in larger groups or as a whole class:
  - Which methods are easiest to understand. Are these also the easiest to do?
  - When using a method without drawing, did you want to draw? Why?
  - Are some methods better for some problems than for others? Why?
  - Were there advantages in using two different methods? If so, what were the advantages?

Assessment
- Did children find effective methods of solving the problems?
- Can children explain their methods clearly?
- Note whether in future problem-solving activities children adopt other methods which have been discussed.

DRAWING IN THE SPACE STRAND

Drawing diagrams and pictures has a place in all the broad areas of mathematics — number, measurement, chance and data — and is particularly important in the space strand. Sketches and accurate geometric drawing are essential for understanding and applying geometric concepts.

Interpreting and making drawings of three-dimensional objects is difficult for many children. Research has shown that pictures or diagrams which adults interpret as three-dimensional are often interpreted by young children as two-dimensional. (Deregowski 1986, quoted in Lovitt & Clarke 1988)

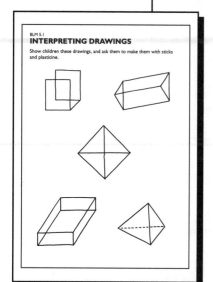

BLM 5.1
INTERPRETING DRAWINGS
Show children these drawings, and ask them to make them with sticks and plasticine.

INTERPRETING DRAWINGS

Years K–3, ages 5–8

• Provide children with sticks and plasticine.
• Show them the pictures on BLM 5.1 (one at a time), and ask them to construct what they see, using the sticks and plasticine.
• Ask children to explain how they approached the task.
• Ask children with different interpretations of the same figure to show their models, and explain what they have done.
• List the methods or strategies, and talk about the different interpretations.
• Ask children to make a different model of one of the figures. This could be a 3D model for a child with an initial 2D interpretation, or a 2D model for a child with an initial 3D interpretation.

Assessment

• Do children interpret the drawings as 2D or 3D?
• Can children make alternative interpretations, after discussion and comparison of their drawings?

DRAWING A 3D OBJECT

Years 3 up, ages 8 up

You will need a sheet of perspex and a set of 3D geometrical models. You can make the models from Geoshapes or Polydron if these are available.

• Set up the perspex vertically between a child and a 3D model, say a cube.

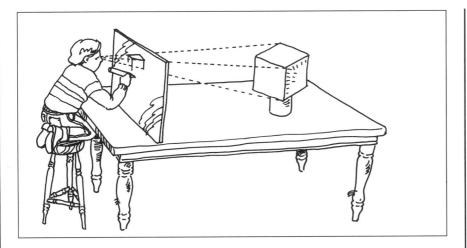

- The child draws the object on the perspex, superimposing the drawn lines directly on what s/he sees. It is important that the child does not move to another position while drawing, as the view will change.
- Ask the children to compare the drawings they make in this way with the illustrations of 3D shapes in textbooks and charts.
- Follow up by asking children to draw the same or other geometric shapes on paper (without the perspex screen), first when the object is visible and later without the object in view.

Assessment
- Do the child's drawings change as a result of this activity?
- Are the drawings more 'realistic' in the way they show a 3D object on a flat 2D page?

Children can draw plane figures and create geometric designs using a variety of tools. Dot paper, squared paper, isometric paper, geoboards, templates, ruler, compasses, protractor and set square can all be used to draw figures such as circles, triangles, rectangles, squares and so on. This material should be available for children to experiment with in an informal way before using it for structured activities.

Using a range of methods enables children to focus on different properties of the figures, or to view those properties from different perspectives.

Computer graphics can also be used. LOGO and Turtle graphics is a particularly valuable program for drawing geometric figures, and learning their properties. Children have to program the computer to draw the shapes they want, and they get feedback from the screen. Does their design look as they had planned it? If not, they can try again. As they do so, they learn, for example, what an angle of 90°, or 60°, or 150° looks like. They learn that the sides of a square are equal, but that four equal sides is not enough to make a square.

Tables

Tables present information of many different kinds, and therefore take different forms. These range from timetables to 'times tables', taking in calendars, timelines, co-ordinate grids, tables of data, and spreadsheets along the way.

It is worrying to learn that 42% of year 6 students in New South Wales were unable to answer correctly this question from the Basic Skills Testing Program (Doig & Masters, 1992).

Sofia lives 83 km from the store. How much will it cost her to have a 3 kg moulding set delivered?

| | free | $2 | $8 |
| | $5 | $10 | $15 |

	Up to 50 km	50 to 100 km	Over 100 km
Under 10 kg	free	$2.00	$8.00
Over 10 kg	$5.00	$10.00	$15.00

A result such as this suggests that more attention needs to be paid to the teaching of tables, particularly given their widespread use in the community and in work situations.

In mathematics, tables can display final results, or be a stage of collecting and organising data in order to solve a problem. Tables can also be used to help to solve problems, particularly combination problems, and classification problems such as the two below.

USING TABLES TO SOLVE PROBLEMS

Years 3–7, ages 8–13
• Pose the two problems. Problem A is a combination problem; Problem B is a classification problem.
Problem A:
How many different ways can you make a dollar using 5c, 10c 20c and 50c coins?
Problem B:
A baker, a butcher and a carpenter all run businesses in the same street. Their names are Pavlova Baker, Hamish Butcher and Ace Carpenter.
Match each person with their business, but:
• no person holds the job that goes with their name
• Pavlova is not the butcher

- Put the children in two groups. One group is to use a table to solve problem A, and then use a different method for problem B. The other group uses a table for problem B, and does not for problem A.
- When children have found solutions, compare and discuss the solutions and their methods.
- How does the table help to solve each problem?
- What are the advantages and disadvantages of using a table compared with other methods?

Assessment

- Can children draw up and use tables effectively?
- Did children find effective alternative methods to solve the problems?
- Can children explain the advantages and disadvantages of using a table?

Children need opportunities to interpret and to make tables of different kinds, in contexts which are of interest and which have purpose. These opportunities will often arise in other areas of the curriculum. In social education, information may be presented in tables or in timelines. In sport or other games, grids can be used to record scores. In science, tables are used to record and display data collected by children. Non-fiction books, newspapers and magazines all provide opportunities for reading and interpreting tables. Even in exploring narratives, children can draw up tables, for example to trace the changes in a character through the story, to make a timeline of the events, or to compare and contrast characters and stereotypes. Some examples of these activities can be found in Johnson & Louis (1985).

TIMETABLES

Timetables are part of children's everyday school life, and opportunities to read and interpret these arise every time a child says 'When do we use the computer?' or 'Which day do we have art?'

Collect local bus and train timetables, as well as more exotic ones, such as airlines, and interstate or overseas trains and buses. Have these available for children to browse through, as well as for more structured activities. Discuss the small print — the legend which tells you that certain buses run on schooldays only, or not on Fridays, or only on market days.

Bus and train timetables can be used when excursions are planned. They can also be used for planning imaginary journeys, for example to plan a trip around Australia, or to plan a holiday in a spot of the child's choice. Children can use timetables to pose problems. The problems they create will be of far greater interest to them than problems out of a textbook or made up by you, and will involve them in more careful study of the timetables.

TIMETABLE PROBLEMS

Years 3–7, ages 8–13

Collect a range of transport timetables, both local and from further afield. You will also need maps of the areas covered by the timetables.

- As a class, work out with the children how you could use (say) a local bus timetable to solve a transport problem. For example

I want to go shopping at Indooroopilly. My shopping will take at least an hour. Then I have to meet my daughter at Queen Street bus station at 5 p.m. We are going to a film which starts at 6 p.m. and finishes about two hours later. Then we go home. How can I do this using buses?

NB Your problem should use places and experiences which are both familiar and interesting to the children.

- Children make suggestions. Write these up, and together work out possible solutions. Discuss these. What are the advantages and disadvantages of each?
- Now ask children to choose from the range of timetables. They should find on a map the area covered by the timetable, and locate the listed stops. They then make up their own problems to give each other.

Assessment

- How successful were children in creating a problem which was realistic and challenging?
- Were the presentations of their problems clear, and could others follow them?
- How successful were children in solving each other's problems, and finding logical and acceptable routes and times?
- Note children who had difficulty accessing the information. Work with them in a small group, using an enlarged copy of the timetable so that all children can see it clearly.

Children can create tables with computers. These may be with word processing packages, or they may use databases and spreadsheets. Spreadsheets are a particularly powerful computational tool, allowing large amounts of data to be processed quickly. Primary school children can use spreadsheets for simple tasks. For example, they can use a spreadsheet to record the heights of the class members, to sort the data, find averages (both mean and median), and to find differences between each child's height and the mean. The details of this will depend on the particular package being used.

GENERAL ACTIVITIES WITH TABLES

All levels, all ages

Choose those activities appropriate for your focus and age group.

• Ask children to find specific information in a table.

• Children can work in pairs to make up questions for others.

• Photocopy a table and white-out some information. Children then complete the table. Children must be able to work out the deleted information logically or from their own knowledge.

• Children can construct timetables for camp, for games, or for special activities in the classroom or school.

• Provide children with a timetables (for buses, trains or planes). Ask them to suggest ways it can be improved for accessing information. (The presentation of most timetables can be improved!)

CO-ORDINATES

Co-ordinate grids are used in many contexts, for example in street directories and maps, games such as Battleships, and in graphs.

Children often first meet co-ordinates that represent squares, as in reading a street directory: B5 is the shorthand for a square on the page. If children have experience of reading tables and other grids such as multiplication and addition tables, they are usually able to interpret and use these co-ordinates without undue difficulty. However, children often have difficulty when moving to using co-ordinates in line graphs, or in latitude and longitude. In line graphs, the co-ordinates represent points on the page so the co-ordinate (4, 5) represents the point found by moving 4 spaces along the horizontal axis, and 5 spaces up.

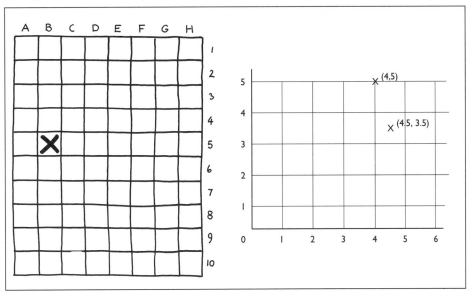

This system enables intermediate points to be located precisely, for example the point (4.5, 3.5).

It is important to be explicit about the difference between these two co-ordinate systems. Too often, the distinctions are blurred, both in the teaching and in the child's mind.

FIND MY HOUSE

Year 4 up, age 9 up

- Demonstrate and discuss with the children the difference between 'street directory' and mathematical 'point' co-ordinates.
- Use a street directory for children to locate the positions of their homes.
- Copy the section (or sections) of the map which shows the children's homes. Instead of labelling the grid with letters across the top and numbers down the side as is usual in a street directory, start in the bottom left hand corner. Label this corner 0, and label each *line* (not square) across and up with a number.
- Now ask the children to mark their house as accurately as they can on the map.
- Using the new co-ordinates, they now describe the position of their house. This will entail the use of decimals.
- Discuss the difference between the two systems.
  - Which is more accurate?
  - Why do compilers of street directories use the first system, although the second one gives a more accurate position?
  - When would you need to use the second system, not the first?

Assessment

- Can children find their house locations in the street directory?
- Can they read the map references in the directory?
- Can children use the new co-ordinate system, giving a reasonable estimate in decimals of the position of their house?
- Can children describe the difference between the two systems?
- Can children describe situations in which the different systems would be more useful?

Computer games will also give children practice in using co-ordinates. For example, 'Maps and Navigation' (designed by Bank Street College of Education) uses latitude and longitude as children find their way, avoid a hurricane, and free a humpback whale.

Completing and creating dot-to-dot pictures is another activity that will provide practice in using co-ordinates.

DOT-TO-DOTS

Years 3–7, ages 8–13

- Using BLM 5.2, children join the points in the order listed.
- Then they create their own picture using the second grid, or a sheet of grid or graph paper. They should use straight lines only. They list the points to be joined, and give their list to another child to complete the dot-to-dot.

Assessment

- Do children mark the points correctly?
- Note children who have difficulty with decimals.
- In their drawings, do children name the points correctly?
- Do children use decimal parts, for example (3.5, 2.2) in creating their own drawings, or do they use whole number pairs such as (3, 2) only?

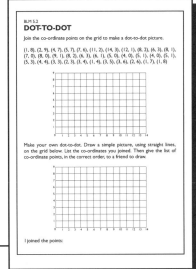

Other activities using co-ordinates are suggested below, in the discussion on graphs. For example, *Graphing the times tables* (p. 101) provides the opportunity to discuss the meanings of the 'in between' points on the lines.

Graphs

Graphs are a way of showing relationships involving time, space and number. Children are usually introduced to picture graphs and bar graphs early in their schooling. These can have a cross curricular focus, particularly in social education where the use of graphs can give the teacher the opportunity to collect information and get to know the children better. For example, graphs can be made about

birthday months	favourite food, hobby
pets	or TV program
eye colour	size of family
hair colour	the countries or cities
how they come to	parents and grand-
school	parents come from

Children can use this kind of information to create a personal profile. The information is also extremely valuable when talking about similarities and differences.

Throughout these activities, the teacher usually prepares the pro forma for the graph, and the children write their names, or paste pictures of their pets, as appropriate, in the correct columns. Some time is then spent discussing the graph, answering questions, and writing statements about it . This kind of modelling is an important part of learning about graphs. However, the graphs which teachers model and demonstrate in these early years are usually bar graphs, so children's experience with different kinds of graphs can be seriously limited.

We need to present different models, even in the early years, so that children become aware that there are many different kinds of graphs, just as there are different kinds of drawing and writing. It is important to encourage children to collect and represent data in ways that make sense to them, just as we encourage them to read and make sense of books in which they may not be able to read every word accurately, and to write from the first days of school, even though they may 'know' how to write only a few letters and perhaps their own name.

In a year 4 classroom, children had considerable experience with graphs, and the teacher believed that they had a sound understanding. They could indeed interpret bar graphs competently, and make accurate statements about the information in them. However, they had never been asked to create a graph from scratch, that is, collect and organise the information and decide for themselves how to present the information in an effective way. They were quite unable to make graphs for themselves, following a reading of *Phoebe and the Hot Water Bottles*. They were asked to draw a graph to show how many hot water bottles Phoebe received each year of her life. (Her father gave her one each Christmas and birthday and whenever she was especially good.) Both teacher and children were frustrated at the impasse reached.

However, the teacher reflected overnight on the situation, and changed her approach. In the next session, she asked the children to imagine something they would like to be given each Christmas, birthday and whenever especially good. Then she asked them to represent the number they would get each year of their life. A wide range of interesting and informative representations were made. By making their own decisions about how to represent the data, and by sharing and comparing these, children learnt much about the advantages and limitations of different graphical representations.

The activity also provided the opportunity for some interesting sociological observations. One group thought you would get more presents in your first year at school, because children are always good when they start school, but that you would get fewer in the following year. One child declared that 'when you were young you would get more presents for being good because little children are good'. A voice of experience from the front row responded to this 'You haven't lived with my little sister!'

We wrote in chapter 3 about the importance of providing a variety of materials and media. Providing squared paper to children who are making graphs is likely to result in them using equal sizes to represent equal quantities (one square = one item). However, it is unlikely to stimulate the understanding that in a graph, equal quantities need to be represented by equal sizes. See how these children have cut out individual hot water bottle shapes for their graph based on *Phoebe and the Hot Water Bottles.*

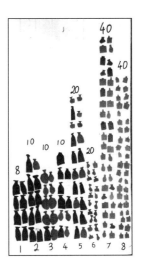

The creation of a graph such as this is a powerful learning opportunity. Pose the questions:

• Can you tell just by looking, not counting, or reading the number, which year has the most hot water bottles?

• Why is it difficult to read?

• What could you have done to make it clearer to the reader?

Because there is often such a strong emphasis on bar graphs, and on a one-to-one representation (that is, one square representing exactly one item) in the early years, it is important when moving into other kinds of graphs to be explicit about the differences in the different kinds of graphs.

Early graphing experiences usually record the number of items in discrete, non-numerical categories, for example pets (dogs, cats, fish …) or colour of cars (red, blue, yellow …). You can also introduce graphs which involve quantities on both axes, by graphing, for example, size of families, or shoe size.

The activity below involves children in collecting, organising and recording data relating two quantities, and is an excellent introduction to notions of variability. Younger children can make the graph, and draw conclusions from it. Older children can extend the activity to investigate statistical concepts such as average and probabilities.

HOW MANY IN A PACKET?

Years 1–7, ages 6–13

Adapt the questions and the follow-up activities to the age and interest of the group of children.

This activity may need two sessions for completion depending on the age of the children and the concepts and skills you decide to focus on.

• Provide a set of miniature packets of sultanas (or Smarties, or matchboxes, etc.), one packet for each child.

• Before children open the packets, ask:

· Do you think each packet has the same number of sultanas in it?

· About how many do you think there will be in each packet?

• Children can count and record the number of sultanas in their packet. You record the number each student has on a class list or on the blackboard.

• Ask children to work in pairs or small groups to organise the information so that it is easier to draw conclusions.

• Share and discuss the way children do this. They may use tables, or lists, probably grouping together students who found the same number. They may order these groups by the number of pieces. Some may spontaneously create a bar graph.

This group recorded the number of sultanas in each packet, and the number of packets that contained the same number. They then made a graph from this information.

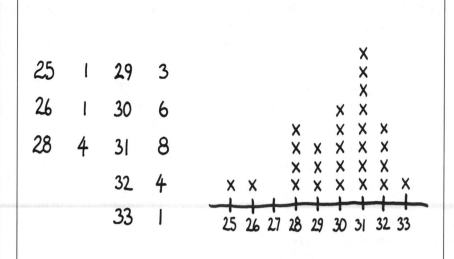

25	1	29	3
26	1	30	6
28	4	31	8
		32	4
		33	1

- Ask all the children to make a graph that shows the information. Highlight any graphs that children have already made, but emphasise that the children should use their own ideas to make a clear graph, as well as adapting ideas they have seen presented.

One form of graph which may arise is shown here:

- Show the children how, once you have some idea of the numbers involved, the data can be recorded straight onto a graph of this kind.

- Discuss:
  - Do the graphs need to start at zero for the numbers of pieces? Why or why not?
  - Should all the numbers between the least number found in a packet and the most found be shown, even if nobody found a particular number? Why or why not?
  - If you opened another packet, about how many would you expect to find in it?
  - What numbers would surprise you very much?

- Children can write statements about the information.

- Older children can discuss assigning numerical values, for example 'The chance of getting 30 or 31 would be about 50%, or half'. 'The chance of getting 27 or less would be about 2 in 28, or 1 in 14.'

  They can design and carry out experiments to test their predictions.

  They can also find the three different kinds of averages: mean, median and mode.

  - In the example given, the mean is found by adding the total number (839) and dividing by the number of packets (28): 29.96.
  - The median is the middle value: there are the same number (of packets) above and below it: 30
  - The mode is the most commonly occurring: 31.

Assessment
- Are children able to count and record the numbers accurately?
- Note the range of ways in which children organise the information.
- Are children able to transfer the information into a bar graph?
- Are children able to discuss probabilities appropriately? For example, after making the graph do younger children predict that there will be about 31 in another packet? Do they understand that there may be a few more or less, but that it is unlikely that there will be more than 34 or fewer than 25?
- Can older children assign numerical values to the probabilities?
- Can they calculate the different averages?
- Can they explain the difference between the mean, median and mode.

GRAPHS AS PICTURES?

There is strong evidence that many secondary school students see graphs as pictures. (See, for example, Swan 1988.) When presented with a graph such as the one below, students describe what is happening (Kerslake 1979):

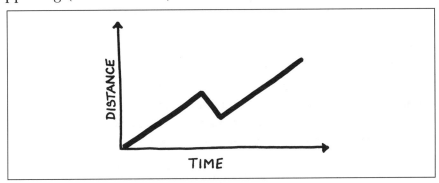

'You're going up a mountain, say, and then coming down a bit, and going up again.'

'He starts at one place, goes round two corners and then on again.'

'You're going NE, then SE and then NW.'

Activities and experiences are needed which challenge this view of a graph as a picture, and which treat explicitly the features of different kinds of graphs.

The distance-time graph in the following activity is an example of a line graph. Every point on the line has a meaning — for example, we can read off the graph how far the person is from the starting point at any time.

A COUNTRY WALK

Years 5–7, ages 10–13

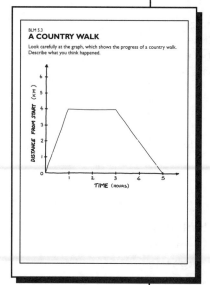

BLM 5.3
A COUNTRY WALK
Look carefully at the graph, which shows the progress of a country walk. Describe what you think happened.

- Present the graph on BLM 5.3. Ask the children to describe what happened on the walk. Keep children's writing for discussion later.
- Walk to the local park for a game or a picnic lunch. Mark regular distances (50 m intervals) with chalk on the footpath or tie a label with the distance on a tree. Children walk in groups of 3 or 4 with a watch or stopwatch. They record the time at each distance marker.
- Record the time spent at the park. Record the time at each distance marker on the return journey.
- Use these times and distances to draw a graph of the distance from school over time.
- Present the original graph again. Compare it with the graphs children have drawn of their own walk.
- Look again at what children wrote about the original graph. Discuss the different interpretations, and any similarities and differences with the actual journey children have undertaken.
- Allow children to modify their interpretations of the original graph if they wish.

Assessment

- Do children interpret the original graph pictorially?
- At the end of the activity, have children revised their interpretation?
- Can children measure distance and time accurately?
- Can children record the time and distance clearly and efficiently?
- Can children transfer the times and distances they have recorded to a graph?
- Note any misuse of conventions in their graphs, for clarification and discussion.

The graph in the following activity is a scattergram. The first part of the activity is also designed to challenge the view of a graph as a picture. Many people reading the graph assume that the vertical (up) axis gives the height of the people, despite the clear label on the horizontal axis.

A scattergram records data on two categories — in this case height and age. It does not tell you the number of people at or between particular heights and ages. A scattergram will show you whether there is a general relationship (or correlation) between the two measures — that is, for example, whether older people are generally taller than younger people.

WHO'S WHO?

Year 3 up, age 7 up

- Children identify the different people in the height/age scattergram on BLM 5.4.

They then fill in an appropriate scale on each axis.

Follow-up activities can include the following:

- Children can create a scattergram for their family or neighbours, based on heights and ages using BLM 5.5. They can fill in their own height and age at school, and complete the graph at home.
- Children devise other scattergrams with class data, e.g. height/weight, height/armspan, neck/wrist, length of foot/height. Some of these may show correlation: that is, the two quantities are related, while some may be more scattered.
- Extend height/age data by measuring height of (say) two children chosen randomly from each class in school. By using children only (no adults), correlation is more likely to be apparent in the graph.
- Ask a question such as 'Do tall people run faster?' Children devise ways of collecting the information, graphing it, and drawing conclusions from it.

Assessment

- Do children misread the graph, to read height vertically rather than as marked along the horizontal axis?
- Can children fill in a reasonable age and height scale on the axes?
- Are children able to create scattergrams with the help of their families?
- Can children recognise in their scattergrams that some quantities are closely related, others are less closely related?
- Note the way children discuss the data and its presentation.
- Can children design experiments and collect and organise data to answer questions relating two measurements, such as 'Do tall people run faster?' or 'Are people with big feet also tall?'?

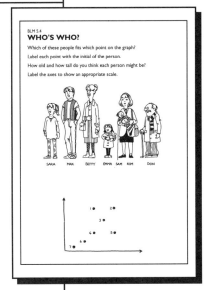

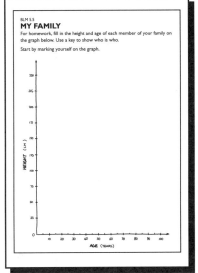

The first part of this activity adapted from *The Language of Functions and Graphs* (Shell Centre 1985).

Information can be expressed in many different ways, for example in tables, graphs, text, or pictorially. The following activity focuses on transferring information from one format to another.

TRANSFERRING INFORMATION

Year 3 up, age 8 up
• Provide children with the following information, showing the temperature in Melbourne through the day.

midnight	9°	noon	23°
2 a.m.	8°	2 p.m.	25°
4 a.m.	8°	4 p.m.	24°
6 a.m.	7°	6 p.m.	16°
8 a.m.	12°	8 p.m.	14°
10 a.m.	16°	10 p.m.	13°

• Use data from your local area instead, if that is available. You can collect this data as a class. Collect readings for the evening, middle of the night and early morning as well through the school day. Children and parents can help with the out of school readings.
• Ask the children, in small groups, to show the information in another form. Each group can choose (or be assigned) one of the following forms.
  • as a table
  • as a graph
  • as text (e.g. a letter to a friend, explaining how the weather affected you, or a report)
  • in pictures
• Discuss:
  • What does each form tell us?
  • What is special about each form?

Assessment
• Are children able to present the information in the range of ways suggested?
• Note any problems children experience with a particular form of presentation.
• Are children able to identify the main features, and the advantages and disadvantages, of each form?

Extensions
• Collect the data over a period of time, and discuss any trends you can observe. Experiment with different ways of displaying the data. For example you can graph the minimum and maximum temperature for each day, or the temperature at a particular time each day. This would give you a better idea of the way the temperature varies from day to day.
• You could record rainfall, or hours of sunshine, and make a scattergram to show whether these were related to temperature.
• Children can design other investigations using weather data, and present their findings in different forms.

GRAPHING RELATIONSHIPS

Graphs show relationships, particularly linear relationships (that is, where one quantity is directly proportional to the other), very clearly. Graphing such relationships from first-hand data is exciting for students, and also provides an excellent basis on which to build algebra concepts and graphs of functions. The two activities below show how children can use their prior knowledge, or experimental data, to graph quantities which are in proportion to each other.

GRAPHING THE TIMES TABLES

Year 3 up, age 8 up
• Ask children to complete a multiplication grid, from 1 to 10.
• Provide graph paper, and ask children to make a graph of (say) the 3 times table.
They show the number of times across the page, and the product (answer) up the page.
They join the points.
• Ask:
  What shape do you get when you join the points?
If any children do not have a straight line, ask them to compare their graph with others. Can they identify what they have done differently?
• You can introduce the algebra notation $y = 3x$, or $y = 3 \times x$. Take care to point out the difference between the multiplication sign x and the letter x!
• Ask children to choose another of the tables and to draw its graph on the same axes (grid).
• Ask children to discuss in pairs the graphs they have drawn.
  What is the same about them and what is different?
• Ask them to predict what they will get when they graph another of the tables. The children then draw the graph they have predicted.
• Now pose the question:
  What do the in between points on the lines you have drawn mean? For example, the point on the 3 x line half way between 3 x 2 and 3 x 3?
  (Reading the in-between points is known as *interpolation*.)

Assessment
• Can children locate the points on the graph accurately?
• For the next table, do children predict another straight line through the origin (0, 0)?
• Can they predict whether it lies above or below the lines they have already drawn?
• Can children identify that the in-between points will give multiplication of, for example, $3 \times 3\frac{1}{2}$?
• If you have introduced the algebra notation, are children able to apply this to another of the tables; for example the four times table is shown by $y = 4x$.

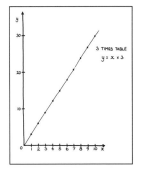

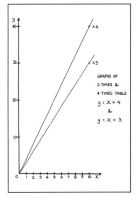

Provide opportunities for children to use graphics packages to create graphs, and to experiment with different formats. Make sure that children have the opportunity to use the computer in small groups. Provide clear tasks, so that time at the computer is not wasted, and make sure the tasks will engage the children in purposeful work and allow for a range of responses. If you are working with upper primary children, try to arrange that at least one group can use the computer when children are collecting and organising data. Compare the results of the computer group with those of the rest of the class.

REFINING THE LEARNING

In the descriptions which follow, we give some idea of the kinds of talk, learning and outcomes which can occur as children share, discuss and refine their graphs. In some cases, this discussion will be at the whole class level. At other times, it may be more appropriate to work with a small group.

The questions for discussion may come from the teacher or from the children. Very often other children will address issues raised by their peers more seriously than they will those raised by teachers. Peer culture can be harnessed in the pursuit of knowledge!

The graphs below are based on the book *10 for Dinner*. This book follows ten children through a party, and tells us the number of children who prefer each song, game, or food, what time they arrive, what they wear, how they travel home, and what presents they bring. This provides simple data which children can organise and analyse, as well as a model for children to follow when collecting their own data.

Holly's graph

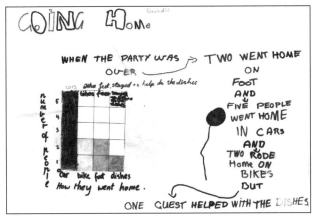

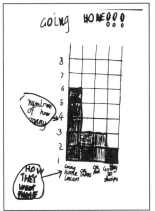

David's graph (left) and Timothy's graph (right).

Holly, David, and Timothy have all represented the ways the children went home from the party in *10 for Dinner*. Some points for discussion might be:

• Compare the number going by car in Timothy's and David's graphs. Which is correct?

• What actually shows one person on each graph?

• Note the zero on David's graph. How could that be useful?

• What has Holly included to make her graph very clear?

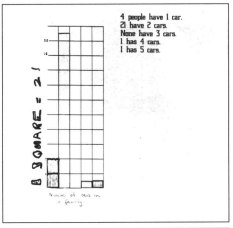

Eye colours (left) and cars (right).

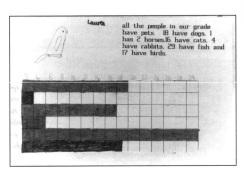

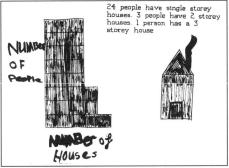

Pets (left) and houses (right).

These children have collected their own data from the class, and made graphs of the information. Some points for discussion might be:

- How have the children shown the larger numbers?
- Which method is easiest to read — the 24 for the single storey houses, 18 for the number of people with blue eyes, 21 for the families with two cars?
- What makes the graph more clear or more confusing?
- How has Lauren shown 17 on her pet graph? And how has she shown 29?
- In the graphs of the cars owned by the families and of the ages of the children, note that a scale '1 square = 2', and '1 square = 2 number of children' has been included. How is this useful?
- Why have the numbers 2, 4, 18 and 20 been written next to the lines in the car graph, instead of opposite the middle of the squares, as in the other graphs?
- Look at the labelling of the axes in these graphs. Which of these graphs are easiest to understand? How does the labelling make the graph more clear?

Maps

From an early age, children are fascinated by maps. One three-year-old we know used to take the road atlas to bed, and trace a route with her finger. Later, children pore over maps in story books, such as the Narnia series. We encounter maps in many different contexts, and use them for many different purposes and the great variety of maps around us should be reflected in the classroom.

Provide a range of maps: sketch maps, strip maps, house plans, the London underground map, other schematic rail maps, street directories, town plans, road maps, atlases, globes.

Ask children to draw maps of many different kinds. They can start from their own knowledge and experience, and the first activity will reveal much about children's perception of space. Young children often draw house plans showing features of importance to them as they imagine themselves moving through the house, but which show little realisation of the way the various spaces within the house are related to each other.

DRAW MY HOUSE PLAN

Year K up, age 5 up
- Ask the children to imagine their own house.
  Start at the front door, walk inside, and move right through the house.
- Ask them to draw a map of their house, showing where all the different rooms and other important features are. Ask them to think about the sizes and shapes of the rooms.

This activity can also be done at home. It is particularly interesting to compare the plans of the same house as drawn by children of different ages, and by adults.

Assessment
• Do children's plans show the relation of the rooms to each other?
• Do their plans show an understanding of shape and relative size?

You can provide experiences that include active involvement, and ask children to map these activities. We have used the book *A Lion in the Night* as a stimulus for a journey around the playground, and a model for children to make a circular map of their own journey.

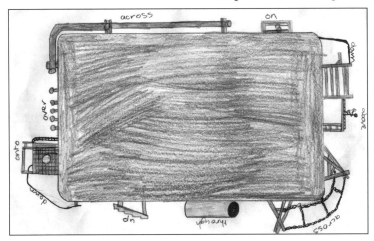

A year 5 child's map of a journey round the playground.

A walk around the school, or around the neighbourhood will provide the opportunity for children to make maps. Because all the children have been involved in the same activity, it offers a rich context for comparison. Ask the children to draw a sketch map first, then walk the route, before revising and refining the map.

A WALK AROUND THE SCHOOL

Year K up, age 5 up
• Ask the children to imagine a walk from the classroom to another spot in the school: the library, the gate, the principal's office or wherever.
• Ask them to close their eyes and imagine they are going on this walk. What do they see as they go? What places or things do they pass on their way?
• They draw a sketch map of their route from memory.
• Ask the children to look at their sketch maps, and mark anything on it which they are not sure about.
• You then walk from the classroom to the other point. Children should observe any differences from their maps, focusing in particular on the areas they are not sure about.
• After the walk, children redraw or refine their maps.

Assessment
- Do children show their route clearly?
- Do children indicate relative distances, sizes and shapes in their maps?
- Do children draw pictures, or use map conventions?

Other activities for mapping within the school are given on pages 49 and 108.

When children are drawing maps of their neighbourhood, for example to show the route from their home to their local shop, each map will show a different route, but many of the maps will have elements in common. This provides opportunities for comparison of the way features are represented, while every child still has an individual task.

Year 3 and 4 children were asked to draw the route from their home to the shops, showing at least six 'points of interest' on the way. After they had completed their first attempt, they were asked to walk around the room, to look at other children's work, before going on to draw another map. As they examined other children's interpretations of the task, many children saw features which they subsequently adopted in their finished map. Some moved from a strip map, showing only the places of interest, to a conventional street map format, showing clearly the relation of the various spaces, and with an indication of the approximate sizes, shapes and distances.

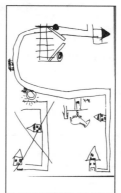
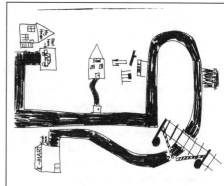

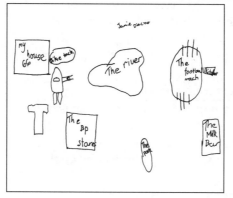
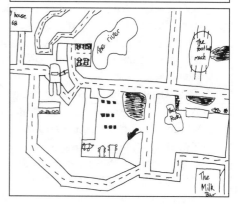

Note the development from isolated features to conventional maps in these children's work.

WALKING TO THE SHOPS

Years 1–7, ages 6–13

• Ask children about their local shop or shops. What is sold there? How far away is it? What do they buy when they go there?

• If it is available, read John Burningham's *The Shopping Basket,* in which Stephen is sent shopping and encounters a number of greedy animals at different landmarks along the route.

• Now ask children to imagine themselves walking to their local shop. They list six landmarks on the way, and then draw a map which shows their house, the landmarks, and the shop.

• When children complete their first attempt, suggest they move around the room, seeing how others have tackled the task.

This is an opportunity for practising their social skills: they should make a point of providing positive feedback to others about their map. For example, they may say 'I like the way you have shown the railway crossing' or 'I like the way you have shown how big the paddock is.'

• Children now complete their maps, adopting or adapting ideas from others.

Assessment

· Do children confidently recall their route from home to shop?

· Do children present a strip map – the landmarks in sequence, but without any attempt at showing distances or the way the various areas relate spatially?

· Do children attempt to show relative distances, and the relation of the various areas passed?

· Do children use map conventions such as a legend, symbols (for railways, roads, etc.), and bird's-eye view?

The following maps were also made by a year 3/4 class. They were involved in a unit of work on trees, and drew maps to show the area of the school grounds where their trees were located. All the maps show an understanding of the space, and the relative positions of the trees, buildings, playground equipment, roads and so on. However, there is an enormous variation in the detail and accuracy, and in the style. Some children have produced picture maps, showing only a few items in profile. Others have drawn a conventional map, showing items in plan, and taking great care to show the sizes of the different spaces involved.

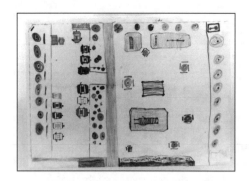

Two maps of the school grounds.

A MAP OF THE SCHOOL GROUNDS

Years 1–7, ages 6–13

This activity may be undertaken in different ways.

- Ask all the children to map the same part of the school ground, or different groups may map different areas,
- Provide a map of the outline only of the school grounds, and ask children to fill in the inside.
- Ask children to make a sketch map before going out to survey the area, or you may wish to take the children on a walk around the area before they start.
- Older children may make scale plans. This will involve measurement of distance (pacing, trundle wheel, tape) and direction (using a compass).
- Children can use colour coding, map symbols and a legend to show the different elements such as buildings, grass, asphalt and so on.

Assessment

- Are the maps clear and easy to interpret?
- Note the ways children represent the different elements in their maps.
- Did children estimate distances reasonably and/or measure distances accurately?
- Could children use a scale correctly?

Making story maps is a powerful strategy for helping children to understand the structure and events of the books they read. The focus here shifts to the understanding of the story; however, making a story map at the same time consolidates understandings and skills in mapping. Story maps involve children in reading and rereading, and in clarifying sequence and place within the story. This strategy can be used from the earliest years, with folk tales such as 'The Billy Goats Gruff', 'Red Riding Hood', and 'Goldilocks', and continue through the primary years.

STORY MAP: RED RIDING HOOD

Years 2–5, ages 7–11

- Tell or read the story, or a number of versions, of 'Red Riding Hood'. Reread the story and list with the children the features they will need for their story map.
  - the mother's house
  - Grandma's house
  - the wood
  - the path through the wood
  - the short cut the wolf took to Grandma's house

  The construction of the story map is linked to the positional and directional vocabulary of the story:
  - the mother's house is on one side of the wood
  - Grandma's house is on the other side of the wood
  - the path goes through the wood
  - the wolf takes a short cut to Grandma's house
- Make sure that the younger children realise what a short cut is. Although they may have read the story many times this is not guaranteed!
- Ask the children to draw a story map using the features of the story and showing:
  - (a) the route Red Riding Hood took through the woods
  - (b) the short cut the wolf took to Grandma's house
- When they have finished they can talk in small groups about their maps.
- Display the maps and talk about the features children have used.

Assessment

- Could the children link the positional and directional vocabulary to constructing a story map?
- How effectively did the children show the short cut?
- Did the children include other features in their maps such as labelling, a legend, the characters?

Many story books contain maps, which help children to extend their understanding of the story, and which also serve as models for children to make their own maps. These maps range from the highly simplified and stylised circular map of *A Lion in the Night*, to the detailed series of maps in Nadia Wheatley's *My Place*.

Older children can use maps for orienteering, learning the use of a compass and angle measures as they do so. The purpose for using the map and the compass, and the physical involvement in the activity, together with the considerable challenge posed to a ten- or twelve-year-old, make this a powerful learning experience.

Maps can be explored together with other mathematical tools, such as timetables (of buses, trains or airlines), tables and graphs (for example of rainfall or temperature). This will provide alternative ways of looking at the same information, and will also provide opportunities for integrating information from different sources.

Activities involving maps are: *Timetable problems*, page 90, and *Find my house*, page 92, *House plan*, page 104, and *Footsteps over the city*, page 49. You may also wish to introduce maps into *A country walk*, page 98, *Maths trail*, page 51 and *The wolf game*, page 17.

Many opportunities for exploring and making maps will occur in different areas of the curriculum. You can use these opportunities to introduce or consolidate particular skills and knowledge, as well as using the mapping to enhance the learning in the particular curriculum topic. An integrated program developed in this way has meaning and purpose for children.

6 | BRINGING IT ALL TOGETHER

This chapter reviews strategies which will contribute to an effective teaching and learning environment and relates these strategies to providing a mathematics program which extends children's use of language. Because we recognise that teachers and schools are working in a range of environments and have different needs, different ways of using this book are suggested. Starting points which will help you in making changes within your classroom are provided, taking into account the needs of different groups.

Many of the activities suggested in the book are also useful for working with staff and parents and provide starting points for discussion, investigation and curriculum change. Classroom and curriculum organisation, and assessment of children's learning, are also discussed briefly, in the context of extending and enriching children's use of language in the mathematics classroom.

The relationship between classroom organisation and curriculum activities is important. It is important to ensure that

- the classroom climate
- the organisation of teaching and learning
- the planning and implementation of the curriculum
- methods of assessment

are inextricably linked to create an integrated learning environment. Although listed separately in order to highlight the importance of each aspect, in practice the four aspects function together.

Classroom climate

If children are to feel confident about offering their ideas and opinions, sharing their recording, reading and writing, and having their ideas examined and challenged by others, the environment in which they work must be one of mutual trust and respect. That is, trust between the teacher and children, between the children in the class, and between parents, teacher and children. Teachers need to continually demonstrate and make explicit to children that the context for learning in the classroom treats all children as important and as learners.

This trust has to be modelled and demonstrated in everyday classroom experiences, as well as in communication between school and home. School communities may consider providing professional development for parents, so that school and home values and educational goals and objectives can be explored together. Parents can learn why and how the school is implementing the mathematics curriculum, based on what we know about how children learn in general, and more specifically how they learn mathematics. Teachers can learn the particular concerns of parents, as well as discovering more of the resources within families, which individual children and the school community may be able to draw on.

Organisation of teaching and learning

Children need to be given opportunities to work individually, in pairs, in groups and as a class to explore and extend the mathematics curriculum and other curriculum areas. The tasks they are given in these learning contexts are important. The roles children play, the responsibilities given to them, and the teacher's expectations will affect the outcomes of the tasks. For example if we believe that children must discuss issues, find their own materials for recording purposes and be involved in group problem solving, then the physical environment needs to reflect this.

Planning and implementing the mathematics curriculum

Language learning in the mathematics classroom has to be active. The learner needs to be actively involved. This means talking, reading, thinking, comparing, writing, listening, reflecting, hypothesising, generalising, evaluating and drawing conclusions. Children are more likely to engage with tasks which demand these skills, than with tasks which focus solely on getting the right answer by manipulating numbers and symbols.

- For example, by reading each other's work **children** will be given:
  - different models on which to base their own writing and explanations
  - opportunities to discuss a variety of approaches to a problem
  - opportunities to discuss more efficient methods of recording and solving problems
  - the awareness that people can approach tasks in a variety of ways
- From children's written work and other recordings, **teachers** will be able to:
  - observe the breadth and depth of work produced by the children
  - select appropriate samples for assessment purposes
  - focus on children with particular needs
  - give feedback to the children and
  - draw the learning together
- By talking and listening to each other **children** will learn:
  - to acknowledge and respect others' opinions
  - to articulate their methods and explanations
  - that there can be a variety of approaches to the same task
  - to appraise and critically challenge others
- As **teachers** listen to children, they can:
  - focus on children's understandings and misconceptions
  - explicitly recognise and value the experience children bring to the tasks
  - explicitly recognise and value the place of spoken language in the mathematics classroom
  - encourage children to be clear and logical in their explanations and presentations
  - encourage respect and support amongst children and at the same time encourage them to challenge each other

Assessment

As you listen to and talk with children as they work, analyse their working notes or their written reports, listen to their oral reports and explanations, you will find evidence of the different purposes for which they are using language.

As children talk, you can listen and observe. You can note significant features of talk, for example, particular insights or misconceptions, or the use of mathematical vocabulary, or mathematical ways of talking. These can then be incorporated into your planning of future curriculum experiences.

You will be able to collect work samples, take notes of spoken language, observe interaction on both a social and cognitive level, all of which will give you information for building up a profile of children's mathematical learning and for reporting on children's achievements.

Much of the information you gather can be shared with students. Giving feedback to students about the mathematics they have used, their explanations, project work and methods of recording gives children clear messages about teacher expectations. This can lead to discussion focused around, for example, choosing efficient and effective methods to use in mathematics. The information will also be valuable for planning future programs, for working with individuals or groups, to redirect children's thinking, to extend and challenge them, or to consolidate their understandings. Using class lists, check lists, observation and listening, work samples, informal and formal testing all contribute to identifying children's learning outcomes and can be used effectively to record and report on children's achievements.

How to get started

Extending the use of language in the mathematics classroom takes time. Not only does it take time in planning and evaluating, it also takes time in the implementation and may affect your current timetable. Moving into such activities may lead you to review how your mathematics program sits in relation to the organisation and time allocation for other curriculum areas.

Children need time to plan, discuss, compare, experiment, evaluate, and reflect. Sometimes this does not appear to be valuable learning time, because children are not directly involved on task. This is one of the differences between the traditional approach and the more open approach discussed in this book. This approach considers mathematics to be much more than learning a number of discrete skills and performing tasks in a specific way.

Throughout this book there are suggestions for a variety of activities which will help you to extend the use of language during mathematics. Some can be used with parents and with other teachers for discussion points and also as starters for looking at your own attitudes and beliefs about mathematics. Because we recognise that there are many different ways to broaden and extend mathematics teaching in classrooms we have suggested different approaches to meet different needs and purposes.

Question: How can I bring some of the issues raised in this book to the attention of the school community, for discussion and planning?

Answer: There are many different starting points to bring issues to the attention of the school community. Here are a few suggestions and you may find one which will suit your own school context.

- Show what you have done to your immediate colleagues at the same grade or department level and get them involved in the activities.
- Collect other documents and reading matter which will support what you are doing. See the bibliography for some useful references. Report on the material to a mathematics committee meeting for discussion, or as a part of examining the mathematics curriculum.
- Suggest that these activities be used as discussion starters.
- Present to parents at a grade night or information night.
- Present the findings of your activities at a staff meeting.
- Suggest that teachers read the relevant chapter and discuss the issues as well as carry out some of the activities.

Question: During mathematics I encourage talk amongst the children but how can I make this more focused rather than having the children move away from the mathematical task as seems to happen sometimes?

Answer: Read chapter 3, which discusses in detail spoken language and active communication. Start with the activity *Finding out* (p. 37). Choose a topic you wish to investigate and list the children's responses according to the headings. The material you gather from this activity will provide you with information for discussion and ideas for planning appropriate experiences around the particular topic.

For a more focused activity to encourage mathematical talk try *Subtraction methods* (p. 40).

Some strategies for keeping small group talk focused and on task are given on pages 41–3. Read and try these.

Remember, talk takes place all the time, so read chapter 4 which focuses on reading and writing in mathematics, where you will find other ideas.

Question: I would like to extend the ways in which children record their mathematics. How can I increase the children's awareness of recording possibilities?

Answer: Read through chapter 2 which deals with recording and notation in a detailed way. Try the activities *Representing division* (p. 13), then *Homes for animals* (p. 18).

If you have children in years K–3 try *Calculator patterns* (p. 20). When you have completed this activity you might like to compare what you believe the children learned and what you have learned with the summary on page 20.

Display the children's work and make sure that time is given for discussing the various methods of recording and the different ways the children may have approached the problems. Make sure you note children who may be having difficulty.

Question: Children in a class have different perspectives on and knowledge about mathematics. How can I find out what the children know about a mathematical topic that I want to introduce in order to support all the children?

Answer: Knowing what children know is important in planning and implementing the curriculum. Use the activity *Find out what your children know* (p. 2). If children work on one or more of the suggestions in groups, you can take time to listen to their opinions and comments. You can then make a class list of their ideas, which will give the children a sense of shared purpose. You may decide to introduce and list other aspects of the topic which they did not raise. Asking children to write individual statements gives you a snapshot of the breadth and depth of knowledge.

You may also like to try the activity *Finding out* (p. 37) which has a more focused approach to finding out what children already know about a mathematical topic.

Question: How do I know whether children are learning? How can I keep track of the different children in my class when they are doing so many different things?

Answer: Assessment of learning is extremely important. As a teacher you need to be able to articulate children's learning to children, parents, other teachers, the administration and in some cases other educational authorities. We have provided assessment indicators with most of the activities in this book, to help you to focus on the particular skills and understandings involved. As you develop your own activities and find or adapt others, you will need also to develop indicators, based on the aims of the activity and on the skills, concepts and understandings involved. Use class lists and checklists, anecdotal records and work samples to help you make informed judgements on children's progress.

Remember that assessment is an ongoing part of your normal classroom work, not an add-on.

If assessment policy in your school does not appear to be in line with the goals of the curriculum, or with styles of teaching and learning which you believe is appropriate and effective, you may need to raise this issue at grade level or department meetings, with the mathematics curriculum committee, or at staff meetings.

Question: How can I organise my classroom in order to optimise the use of language?

Answer: Start with planning. You can use the grid on BLM 6.1 to help you to include all the communication modes as you work in a mathematical topic or in a cross-curriculum topic.

Following are some practical suggestions for organising the classroom:

- Make sure that the materials and resources you will need are readily accessible not only for you but for the children. These materials will include a variety of paper and writing tools, measurement instruments, calculators, a range of concrete materials, books, newspapers, magazines and so on.

- Set up the classroom so that small group work and discussion are facilitated. For example, arrange tables or desks so that children can face each other.

- Provide a balance of individual, small group and whole class work. Organise your time so that you interact with every child on a regular basis, individually and in a small group. Use a class list to plan and record your observations and interactions.

- Spend some time establishing co-operative working skills such as listening, checking, encouraging and focusing on the task. (See page 43.)

- Provide children with opportunities to share and discuss their work.

BLM 6.1
PLANNING FOR ALL LANGUAGE MODES
Topic:

ACTIVITIES	SPOKEN Listening/speaking	WRITTEN Reading/writing	ACTIVE Observing/doing	GRAPHIC Interpreting/ drawing, making
Finding Out	Discussion	Listing		
Decimal Stories	Explanation	Writing		
Estimating Distance	Discussion		Measurement outdoors	
Junk Catalogues (adding money)	Discussion	Listing		
Decimal number line				Drawing interpreting
Guess my number (0-10, 1 decimal place)	Playing a game			

Comments:
Need to involve children. Look through list of hobbies for other topics.

Question: I believe that extending children's language use will help children to learn more effectively. However, the parents at this school believe that mathematics should be taught in traditional ways. How can I make changes and retain their support?

Answer: Getting parents' support is important. You will need to start in small ways and take parents into your confidence as you further implement changes

Make sure the children are aware that when they are talking about, acting out, reading, drawing, or writing about mathematics and mathematics activities they are 'doing' mathematics. Some children, when they are doing anything other than manipulating numbers, do not view the activity as mathematics. This may cause concern when a parent asks the inevitable question, 'What did you do in maths?' and the answer is 'We didn't do any maths', when in fact the children have been involved in reading, writing and talking in mathematics.

So if the children are writing mathematics journals, or searching the newspaper for mathematics (see page 55), or arguing about the best method to use for subtraction (see page 40), ensure that they realise that what they are doing is mathematics! The child who can explain that, for example, they are not 'just reading and writing' or 'just talking about it' but learning about mathematics and practising specific mathematics skills as they are involved in different activities will be a powerful advocate for the changes you are making.

Explain to parents what you are doing, and why you are making changes. Most families will be very supportive, once they understand what you are doing. At your grade meeting evening or at a parent information evening try these activities.

- Give parents the short newspaper extract on BLM 4.1 and ask them to identify and record the mathematical understandings needed to get the meaning from the article. Start them off working individually and then ask them to compare their findings in groups.
- Ask parents to work in pairs to complete the activity *Magic squares* (p. 58).
- Try the activity *Number cards* (p. 47) or *Maths walk* (p. 48). Both these activities involve physical movement.

Have someone observe the parents at work and build up a list of the different kinds of language they use.

These three activities include the four language modes; reading, writing, spoken language and active communication. Computation, knowledge of mathematical terminology, solving problems, spatial relations and working co-operatively are essential for participation in these activities. You may choose other activities, but make sure they demonstrate the importance of language use, including active communication. Do not forget to include activities which involve computation when you talk to parents, so that they realise that their children are learning number skills as they use the language modes to extend their mathematics experience.

You could also have some of the children's responses to the same activities on display so the parents can see how the children worked. This will give you an ideal opportunity to discuss the ways the children approached the problems. Because the parents have been involved in the same activities, they will be able to contribute to the discussion.

Question: I am a student teacher looking for activities for my teaching practice. What could I undertake that will take just one or two lessons?

Answer: You could collect some information about the children with whom you are working. For example, use the activity *Decimal stories* (p. 14) to find out children's perceptions of decimals. You can adapt the activity to another topic, for example Subtraction Stories or Multiplication Stories. (See Griffiths & Clyne, 1994.)

For one or two lessons you can try *Calculator patterns* (p. 20), the *Wolf game* (p. 17), or *Number cards* (p. 47) with younger children.

These provide opportunities for writing, recording, reading a story, acting out, and physical involvement.

With slightly older children, try *Circumference of a circle* (p. 50), *Riddle book*, (p. 22) or *Vets in Australia* (p. 60). These provide opportunities for active involvement, investigating vocabulary, and building on children's knowledge.

These will help you to get started, and are easy to implement in the difficult situation of working in someone else's class.

Question: I have been trying to involve my children in activities that use the language modes and have collected quite a folio of work which include examples of all the modes including the active communication. How can I approach my colleagues with the idea of extending language use, using my folio and the material in this book to help?

Answer: Having work samples and a number of anecdotes to show other teachers is a sure way of gaining other teachers' attention in any learning area. At departmental or grade level meetings, show your colleagues the work you and the children have been doing and suggest some joint planning, perhaps beginning a topic, finding out what children know, or using calculators.

Working with another teacher in his/her classroom is a highly effective way to demonstrate the approach and to initiate discussion about teaching and learning. Try combining two classes for one or more sessions in order to try out together some of the ideas in this book.

Question: As professional development co-ordinator, how can I use this book to help my colleagues to broaden and extend their language use when working in mathematics?

Answer: Find activities such as *Finding out* (p. 37), *Junk mail catalogues* (p. 64), *Calculator patterns* (p. 20), and *Making polygons* (p. 49), which can be used at a range of age levels. Use these yourself, and ask other teachers to try them. At a staff meeting, display work samples, and discuss what happened during the activities as well as talking about what they reveal about children's learning of mathematics.

Involve the staff in an activity such as *Generating discussion on mathematics and language* (p. 3) or *Linking mathematics and reading* (p. 55), as a starting point for discussion. Or ask teachers to do one of the activities for children **before** showing and discussing the children's work. This will make an interesting starting point for comparison and discussion.

Ask a small group to investigate the issues raised for discussion at the next meeting. Be sure to include in the group a cross-section of the staff.

Set a date for further discussion, and ask for some commitment from your colleagues, whether it be to try activities in the classroom, or to read and reflect on the issues, or read a chapter for discussion before the next meeting.

When some activities have been trialled you may like to suggest an evening for parents to view some of the children's work, do some of the activities and become more informed about their children's learning.

Conclusion

Extending and broadening the language in your mathematics program will not happen overnight. If you are already doing this, your task may be to make these changes as effective and as integrated as possible, by enlisting the support of the whole school community — children, teachers, administrators and parents. This will be gained by:

- demonstrating the effectiveness of the approach in actual classroom practice
- talking about the issues, in both formal and informal situations
- involving others in activities which embody the principles of the approach
- working with others to plan and to implement the approach

Whether you are a beginning or an experienced class teacher, a student teacher, or a professional development co-ordinator, this book will help you to extend and enhance the use of language in the mathematics classroom.

By doing this you will be giving your mathematics program a much broader context for learning, making your program more relevant and meaningful, and extending children's experience, uses and understanding of mathematics.

WHAT DO YOU KNOW ABOUT _____ ?

Write and draw what you know about _____

Think about these questions:

  • How do people use it? _____

  • Where would you find it? _____

Think of as many different examples of _____ as you can.

What else do you know about_____ ?

Find a partner.
Talk about what you both know.
Choose two pieces of information to tell the others in the class.

THE WOLF AND THE SEVEN LITTLE KIDS

A mother goat lived in a house with her seven little kids. She knew a hungry wolf lived nearby so she looked after the kids every minute of the day.

One day the mother goat had to go out. She warned her children not to open the door to anyone except herself, and especially not to the big bad wolf.

'How will we know if it's the wolf?' asked the eldest of the kids.

'You will know him by his deep growly voice and his black paws,' said the mother goat. 'My voice is high and sweet, and my feet are white.'

When the mother goat had left the house, the wolf came along and knocked at the door.

'Let me in,' he growled.

'Oh no,' said the second little kid. 'We can only let in our mother. She has a sweet, high voice, not a deep, growly voice. You are the big bad wolf.'

So the wolf went away, angry.

He went to a beekeeper who lived nearby, and said 'Give me some honey, or I will eat you up.'

The beekeeper gave the wolf a large jar of honey, which the wolf gobbled down. It made his voice sweet and high. So he returned to the goats' house.

'Let me in,' he said, in his new sweet, high voice.

The third little kid was about to open the door, when the fourth little kid said 'Wait! Show us your feet.'

So the wolf put his feet up on the windowsill, where the little kids could just see them.

'Oh no!' said the fifth little kid. 'You are not our mother. You have black feet, and she has white feet. You are the big bad wolf.'

So the wolf went away again, angrier still.

He went to the flour mill, and said to the miller 'Give me some white flour, or I will eat you up.'

The miller gave the wolf a bag of flour, which the wolf used to dust his feet, and turn them white, before he returned to the goats' house.

'Let me in,' he said, in his sweet, high voice. 'I am your mother. See my nice white feet.'

He put his feet up on the windowsill to show the little kids.

'Yes, I'll let you in mother,' said the sixth little kid, and opened the door.

What a commotion as the little kids dived for cover and the wolf chased them!

He caught the first little kid under the kitchen table, and swallowed him whole.

He caught the second little kid behind the dresser, and swallowed her whole.

He caught the third little kid between the bed and the wardrobe, and swallowed him whole.

He caught the fourth little kid inside the wardrobe, and swallowed her whole.

He caught the fifth little kid on top of the bookshelf, and swallowed her whole.

He caught the sixth little kid up the chimney, and swallowed him whole.

The seventh little kid hid in the grandfather clock, and the wolf could not find her.

So the wolf staggered off, full with the six little kids.

Soon the mother goat came home, and found the door wide open, and the house empty.

She called for her children. A little voice came from inside the grandfather clock. 'Here I am, mother. The wolf came and ate all my brothers and sisters.'

The mother goat collected her big scissors, needle and thread, and set off with the seventh little kid. They followed the wolf's footprints.

Soon they caught up to the wolf, who had fallen fast asleep on his back under a big tree.

The mother goat took out her scissors, and cut a slit in the wolf's stomach. The wolf was so fast asleep after his greedy meal that he did not wake up.

Out jumped the sixth little kid, alive and well.

Out jumped the fifth little kid, alive and well.

Out jumped the fourth little kid, alive and well.

Out jumped the third little kid, alive and well.

Out jumped the second little kid, alive and well.

Out jumped the first little kid, alive and well.

'Fetch me a large stone,' said the mother goat, and the seventh little kid ran quickly to find one.

The mother goat placed the stone in the hole she had cut in the wolf's stomach, and sewed up the slit.

Then the goats all ran home and never opened the door again when their mother was out.

NUMBER WORDS

Many words relating to number come from Latin and Greek number words. Look for the following elements:

	Latin	Greek	Other
1	*uni-* as in unicorn	*mono-* as in monocycle	words based on *one*, such as alone, once, only some words for first starting with *pr-*, such as prince, prime
2	*bi-* as in biplane *du-* as in duet	*di-* as in dilemma	words based on *two*, such as twice, between, twig
3	*tri-* as in triangle	also *tri-* as in tripod	
4	*quadra-* as in quadrangle, *quart-* as in quartet	*tetra-* as in tetrahedron	
5	*quintus-* as in quintuplet	*penta-* as in pentagon	
6	*sexta-* as in sextant	*hexa-* as in hexameter	
7	*septem-* as in September	*hepta-* as in heptathlon	
8	*oct-* as in octave	*octo-* as in octopus	
9	*novem-* as in November	*nonus-* as in nonagon	
10	*deci-* as in decimal	*deca-* as in decathlon	words such as dime, tithe, meaning *one tenth*
100	*centi-* as in centigrade	*hecto-* as in hectare	
1000	*milli-* as in millennium	*kilo-* as in kilometre	

WHAT ORDER?

Cut out the cards, and put each set in order.

Place in order of price:

expensive toy car	expensive sports car	cheap car

Place in order of distance:

long walk	long car drive	short car drive

Place in order of height:

tall woman	tall doll	small tree

Place in order of weight:

heavy baby	light van	heavy man

MAKE A VENN DIAGRAM

Who plays tennis AND basketball?
Who plays tennis OR basketball?

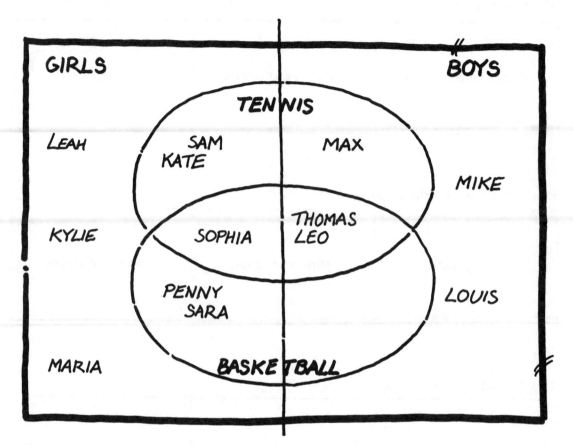

Make your own Venn diagram.
Use your own ideas for the information to record in it.

NUMBERS IN OTHER LANGUAGES

Our class is investigating the ways numbers are said and written in other languages. Please help us by completing the table below for your home language.

English		Language:	
Numeral	Name	Numeral	Name
1	one		
2	two		
3	three		
4	four		
5	five		
6	six		
7	seven		
8	eight		
9	nine		
10	ten		
11	eleven		
12	twelve		
13	thirteen		
14	fourteen		
15	fifteen		
16	sixteen		
17	seventeen		
18	eighteen		
19	nineteen		
20	twenty		
21	twenty-one		
30	thirty		
40	forty		
50	fifty		
60	sixty		
70	seventy		
80	eighty		
90	ninety		
100	one hundred		
1 000	one thousand		
10 000	ten thousand		
100 000	one hundred thousand		
1 000 000	one million		

Please include any other interesting information about number names in your language. For example, are numbers written from left to right (as in English), right to left, or down the page?

NUMBER CHARTS

ARABIC

١	٢	٣	٤	٥	٦	٧	٨	٩	١٠
١١	١٢	١٣	١٤	١٥	١٦	١٧	١٨	١٩	٢٠
٢١	٢٢	٢٣	٢٤	٢٥	٢٦	٢٧	٢٨	٢٩	٣٠
٣١	٣٢	٣٣	٣٤	٣٥	٣٦	٣٧	٣٨	٣٩	٤٠
٤١	٤٢	٤٣	٤٤	٤٥	٤٦	٤٧	٤٨	٤٩	٥٠
٥١	٥٢	٥٣	٥٤	٥٥	٥٦	٥٧	٥٨	٥٩	٦٠
٦١	٦٢	٦٣	٦٤	٦٥	٦٦	٦٧	٦٨	٦٩	٧٠
٧١	٧٢	٧٣	٧٤	٧٥	٧٦	٧٧	٧٨	٧٩	٨٠
٨١	٨٢	٨٣	٨٤	٨٥	٨٦	٨٧	٨٨	٨٩	٩٠
٩١	٩٢	٩٣	٩٤	٩٥	٩٦	٩٧	٩٨	٩٩	١٠٠

CAMBODIAN

១	២	៣	៤	៥	៦	៧	៨	៩	១០
១១	១២	១៣	១៤	១៥	១៦	១៧	១៨	១៩	២០
២១	២២	២៣	២៤	២៥	២៦	២៧	២៨	២៩	៣០
៣១	៣២	៣៣	៣៤	៣៥	៣៦	៣៧	៣៨	៣៩	៤០
៤១	៤២	៤៣	៤៤	៤៥	៤៦	៤៧	៤៨	៤៩	៥០
៥១	៥២	៥៣	៥៤	៥៥	៥៦	៥៧	៥៨	៥៩	៦០
៦១	៦២	៦៣	៦៤	៦៥	៦៦	៦៧	៦៨	៦៩	៧០
៧១	៧២	៧៣	៧៤	៧៥	៧៦	៧៧	៧៨	៧៩	៨០
៨១	៨២	៨៣	៨៤	៨៥	៨៦	៨៧	៨៨	៨៩	៩០
៩១	៩២	៩៣	៩៤	៩៥	៩៦	៩៧	៩៨	៩៩	១០០

CHINESE

一	二	三	四	五	六	七	八	九	十
十一	十二	十三	十四	十五	十六	十七	十八	十九	二十
二十一	二十二	二十三	二十四	二十五	二十六	二十七	二十八	二十九	三十
三十一	三十二	三十三	三十四	三十五	三十六	三十七	三十八	三十九	四十
四十一	四十二	四十三	四十四	四十五	四十六	四十七	四十八	四十九	五十
五十一	五十二	五十三	五十四	五十五	五十六	五十七	五十八	五十九	六十
六十一	六十二	六十三	六十四	六十五	六十六	六十七	六十八	六十九	七十
七十一	七十二	七十三	七十四	七十五	七十六	七十七	七十八	七十九	八十
八十一	八十二	八十三	八十四	八十五	八十六	八十七	八十八	八十九	九十
九十一	九十二	九十三	九十四	九十五	九十六	九十七	九十八	九十九	百

ESTIMATING DISTANCE

You will need a metre ruler or a stick one metre long.

At school

Think of a distance at your home, for example, the length of the garden path, of your bedroom, of the garage.

I am thinking about the length of _____

I estimate the length to be _____

At home

Look at the distance you have been thinking of. Make another estimate of its length.

My second estimate is _____

Look at your metre stick or ruler. Now make a third estimate.

My third estimate is _____

Now measure the actual distance using your metre stick or a tape measure.

The actual length is _____

Write about the activity. What helped you to make your first estimate? Your later estimates? What would you use next time to help you estimate another distance?

THE KING WHO WAS TIRED OF WAR

Once there was a king who was tired of war. He longed for another occupation, which would be just as interesting and challenging, but which would not mean death and destruction.

He therefore offered a reward to the person who could find him an occupation as absorbing and challenging as leading his armies into battle.

A scholar named Sessa invented the game of chess and brought it to the king. The king was delighted with the game, its complexity and subtlety.

He asked Sessa to choose any reward.

'Thank you, your Majesty,' said Sessa. 'All I wish is this: as many grains of corn as it would take to fill the sixty four squares of this chessboard, putting one grain on the first square, two on the next, then four, eight and so on, doubling each time.'

'Surely,' said the king, 'the game is worth a better reward than that! Think again.'

'No,' said Sessa, 'I am a modest man, and that will satisfy me.'

So the king called his vizier to calculate the amount of corn needed, and to bring the bag of corn to Sessa.

Away went the vizier, to set his reckoners to work on the calculation.

STOP HERE. HOW MANY GRAINS OF CORN DO YOU THINK WERE NEEDED ON THE LAST SQUARE? HOW MANY ALTOGETHER?

It was hours later that the vizier appeared again before the king.

'Well,' said the king, 'have you the bag of corn which Sessa has requested for his reward?'

'Your Majesty,' said the vizier, 'it is not possible. All the grain in your kingdom will not be enough. Even all the granaries of the world do not hold the amount you have promised.'

'How can that be? How many grains are needed?'

'Your Majesty, the last square of the chessboard would need 9,223,372,036,854,775,808 grains of corn, and the total number of grains on the board would be 18,446,744,073,709,551,615.

'What's to be done?' asked the king. 'Since I cannot pay this reward, what do you advise me to do?'

'There is one way,' replied the vizier. 'Tell Sessa he can have his reward, if he counts the grains of corn himself. In his whole lifetime, working day and night, he would be unable to count but a small fraction of the number of grains before he died.'

SANDWICH SALES

Look carefully at the graph, which shows the number of sandwiches sold at a school canteen one week.

Write an explanation for the different numbers of sandwiches sold on the different days.

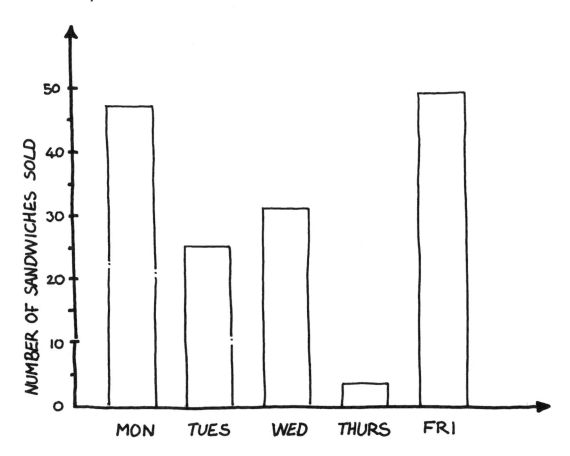

CUPS GAME

For homework, your child has been asked to play a number game with you. Let him/her teach you the game, which is called **Cups**.

In this game, your child will be practising addition and subtraction.

When you have played the game a few times, you may like to discuss with your child what is happening in the game. The following questions may be useful.
• How do you know what number to guess?
• What do you see in your mind when you are guessing?
• How would you guess if we had more counters in the game – say 20?

Here are the rules. Let your child explain them.

HOW TO PLAY CUPS

You will need a set of between 5 and 10 counters (bottle tops, buttons, paper clips etc.) and a cup.

Count the counters, and make sure you agree on the total number (say, 7).

One player closes his/her eyes, while the other player hides some of the counters under the cup.

The first player has to 'guess' how many are under the cup.

Take turns at hiding and guessing.

EVENS AND ODDS

For homework, your child has been asked to play a number game with you. Let him/her teach you the game, which is called **Evens and Odds**.

In this game, your child will be practising addition and learning about even and odd numbers.

When you have played the game a few times, you may like to discuss with your child what is happening in the game. The following questions may be useful.

• How can you tell whether you have an even or an odd number?

• Is it better to be even or odd, or doesn't it matter?

• If I put up an even number and you put up an odd number, who will win? Can you tell without counting?

Here are the rules. Let your child explain them.

HOW TO PLAY EVENS AND ODDS

One player is 'odd', the other 'even'.

Both players hide their hands behind their backs.

At a count of three, both players hold their hands in front of them, with some fingers up and some curled down.

Count the total number of fingers up on each hand.

If the total is odd, the 'odd' player wins. If the total is even, the 'even' player wins.

If you want to score, the total each time will be the winner's score.

NOUGHTS AND CROSSES EXTRA!

For homework, your child has been asked to play **Noughts and Crosses** with you. But not just the ordinary Noughts and Crosses, some new versions. Let him/her teach you the games. Play the ordinary version first.

When you have played the game a few times, you may like to discuss with your child what is happening in the game. The following questions may be useful to start discussion.

• What's the same about this version and the ordinary version?

• What's different?

• Why did you try that move?

• I wonder what will happen if I try this move. What do you think?

• Is there any other game that is a bit like this one?

• How can I stop you winning? (or you stop me winning)

Here are the rules. Let your child explain them.

HOW TO PLAY NOUGHTS AND CROSSES EXTRA

All the games are played on the normal Noughts and Crosses 3 x 3 grid.

Noughts and Crosses: One player puts down Xs, the other Os, in turn. The object is to make a line of three of your mark, in a row, column or diagonal.

Your Choice Noughts and Crosses: In this version, you still take turns, but you may put down either an X or a O, and you don't have to make the same mark each time. The winner is the one who completes a line of 3 the same, Xs or Os.

Last One Wins: You are allowed to mark as many spaces as you like, as long as they are all in the same row or column (not diagonal). The player who fills in the last space is the winner.

Number Noughts and Crosses: In this version you use the numbers from 1 to 9 instead of Xs and Os. Each number can only be used once in a game. Take turns to write a number in a space. The winner is the one who completes a row, column or diagonal that adds up to 15.

LEAD LEVELS WORRY DOCTORS

Melbourne: Children in Australian cities have two to five times the blood lead levels of American children because of more lead in our petrol, a doctor said yesterday.

Community pediatrician Dr Garth Alperstein said most children living in Australian cities had lead levels between 10 and 20 micrograms per decilitre of blood.

"On a population basis that's concerning — we can do a lot better," he said.

"In the US, preschoolers in the 1970s had an average level of 15, but that's now down to four, mainly as a result of reducing lead in petrol."

Above 10 is the new United States' level of concern.

Dr Alperstein works for the Central Sydney Community Health Services and Royal Alexandra Hospital for Children. He and six other medical experts said in *The Medical Journal of Australia* that plans were urgently needed to do something about environmental lead.

The doctors found blood lead levels of children living in an area near a lead smelter were almost identical to those of children in other areas whose main exposure was lead in petrol and paint.

The doctors said it would be worthwhile for the petroleum industry to make a public commitment to reduce lead concentration in leaded petrol by mid-1994 to that of most European countries — 0.15 grams a litre.

The *Courier-Mail*, Brisbane, Monday 5 April 1993

MAGIC SQUARES

Long ago in China, the Emperor Yu was standing by the Yellow River, thinking of how best he could rule the land and bring peace and prosperity to the people. Suddenly, there appeared before him a tortoise. But this was no ordinary tortoise. Instead of the normal pattern of lines, this tortoise bore a pattern of mystic symbols on its back.

Realising that this was a divine tortoise, the Emperor observed the pattern in detail. As suddenly as it had appeared, the tortoise disappeared. The Emperor hurried back to his palace, and called for brush and ink. He drew the pattern of mystic symbols which he had seen on the tortoise's back.

He called for his advisers: astronomers, astrologers, mathematicians, poets, priests, to tell the meaning of this apparition and of the message on the tortoise's back. All agreed that the tortoise was divine, and the pattern was indeed magic. It was named the lo-shu. Since that day, the lo-shu or magic square, has been used in China for fortune-telling, and as a charm against disease.

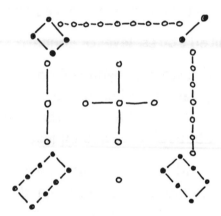

QUESTIONS

- Write the numbers shown by the dots of the lo-shu.
- Look at the lo-shu or magic square. What patterns can you find?
- Make a magic square of your own. Compare your square with the lo-shu and with others made by your class.
- Did you use the numbers 1 to 9? If so, what is the same as and what is different from the lo-shu?
- Did you use different numbers? If so, what was the magic number for your square?
- How did you make the square? Compare your methods with others in your class.
- Magic squares can be of different sizes. The lo-shu is a 3 x 3 square, and is called a magic square of order 3.
- Can you make a magic square of order 2 (size 2 x 2)?

RABBITS!

A lifestyle that's geared to growth

As the breeding season approaches, the female rabbit (doe) joins a small group of up to three males (bucks) and up to seven other does in a well-defined territory. Once pregnant, she carries the foetuses for about 30 days.

In the last week of pregnancy she digs a special breeding hollow, or stop, in which to give birth. This is usually a domed chamber at the end of a tunnel. If her status is high in the group, the stop will be deep in the warren. There her kittens will be safer from predators than those of does lower on the social scale who have to dig stops where they can.

A few hours before the birth, the doe works feverishly to line the stop with dry grass. An hour before the birth she plucks fur from her belly and thighs and uses it to cover her naked, newborn kittens. The birth takes five minutes, after which she emerges from the stop and seals the entrance with soil. She may now immediately mate and become pregnant again.

The doe visits her kittens once or twice a night for no more than five minutes at a time. But because rabbit's milk is richer than both cow's or goat's milk, this is enough to provide them with nutrition. At three weeks the kittens emerge to eat grass but may still approach their mother to suckle occasionally.

Parental concern doesn't last long. If the doe is pregnant, she must prepare for her new litter and, within a few days, she drives her kittens away.

The wild rabbit's breeding capacity is legendary. In harsh environments, like northern Europe, this counters the high death rate — sometimes as high as 90 per cent of the young — caused by predators, climate and disease.

Rabbits breed in winter and spring when green feed is available. There is a strict hierarchy among both sexes in the breeding groups, status being won in fights and displays of aggression. The dominant bucks and does protect the group's territory, which they mark out with scent.

The dominant does, about 25 per cent of the females, produce more than half the young. The second and biggest group (about 43 per cent) produces all but seven per cent of the rest.

The doe can become pregnant at three months. Litters vary from three to seven kittens and she may have as many as nine litters per season — though five or six is the average. In normal conditions she may produce 25 young in a season. Before the season ends, does from her earlier litters may also be producing.

Rabbits, which generally live for two to three years, can go without drinking as long as their food contains at least 55 per cent water. Green vegetation usually has between 70 and 80 per cent. Dry plants have between five and 25 per cent and when there is nothing else, rabbits die.

Rabbits stop plants from regenerating by eating their seedlings. In parts of Australia the result is that various plant species, particularly the woodland trees and tall shrubs, are destined for extinction. In parts of north-east South Australia, even in areas where livestock is absent, cassia, mulga and other acacias are seriously threatened and have died out in places.

TO BRACKET OR NOT TO BRACKET (I)

Place the cards (see BLM 4.5) in the correct positions on the board. Some have been done for you.

Discuss with your group the right place for each of the cards.

Make sure everyone in your group agrees.

WORDS	SYMBOLS	SYMBOLS & BRACKETS	ANSWER
	8 - 3 - 2		
I had 8 apples. I gave 3 to Joe but he gave me 2 back. How many did I have then?			
		8 + (3 - 2)	
			13

TO BRACKET OR NOT TO BRACKET (2)

Copy onto thin card for best results.

Cut out and shuffle the cards. Give them to a small group, together with the board on BLM 4.4. The children have to place the cards in the correct places on the board.

I had 8 apples. I gave 3 to Joe and 2 to Marina. How many did I have then?	$8 - 3 - 2$	$8 - (3 + 2)$	3
I had 8 apples. I gave 3 to Joe but he gave me 2 back. How many did I have then?	$8 - 3 + 2$	$8 - (3 - 2)$	7
I had 8 apples. Joe gave me 3, and I gave 2 to Marina. How many did I have then?	$8 + 3 - 2$	$8 + (3 - 2)$	9
I had 8 apples. Joe gave me 3, and Marina gave me 2. How many did I have then?	$8 + 3 + 2$	$8 + (3 + 2)$	13

READING AND WRITING DIVISION (1)

Place the cards (see BLM 4.7) in the correct positions on the board.
Some have been done for you.
Discuss with your group the right place for each of the cards.
Make sure everyone in your group agrees.

STORY PROBLEM	WORDS	÷	ANSWER	)‾	ANSWER
	8 divided by 2				
2 buns are shared between 8 children. How many buns does each child get?			$\frac{1}{4}$		
				6)12	
				12)6	
		$4 \div \frac{1}{2}$			
		$\frac{1}{2} \div 4$			

READING AND WRITING DIVISION (2)

Copy onto thin card for best results.
Cut out and shuffle the cards. Give them to a small group together with the board on BLM 4.6. The children have to place the cards in the correct places on the board.

Listen to the children as they make decisions.

You will learn about

• their understanding of division and of fractions

• the ways they interact in a problem solving situation

• the ways they use language to help them solve problems.

8 buns are shared between 2 children. How many buns does each child get?	8 divided by 2	$8 \div 2$	4	$2\overline{)8}$	4
2 buns are shared between 8 children. How many buns does each child get?	2 divided by 8	$2 \div 8$	$\frac{1}{4}$	$8\overline{)2}$	$\frac{1}{4}$
You have $12. Each box of chocolates costs $6. How many boxes can you buy?	12 divided by 6	$12 \div 6$	2	$6\overline{)12}$	2
You have $6. A kilogram of nuts costs $12. How many kilograms can you buy?	6 divided by 12	$6 \div 12$	$\frac{1}{2}$	$12\overline{)6}$	$\frac{1}{2}$
Four kilometres of racing track is split into half kilometre sections. How many sections are there?	4 divided by $^1/_2$	$4 \div \frac{1}{2}$	8	$\frac{1}{2}\overline{)4}$	8
Half a kilometre is split into 4 sections. How long is each section?	$^1/_2$ divided by 4	$\frac{1}{2} \div 4$	$\frac{1}{8}$	$4\overline{)\frac{1}{2}}$	$\frac{1}{8}$

MONITORING WRITING IN MATHEMATICS

Checklist	MON	TUES	WED	THUR	FRI
Summaries					
Translations					
Definitions					
Reports					
Personal writing					
Labels					
Instructions					
Notes					
Lists					
Evaluations					
Descriptions					
Predictions					
Arguments					
Explanations					

Over a week, record the different kinds of writing children use in mathematics. Note down any comments and plans.

INTERPRETING DRAWINGS

Show children these drawings, and ask them to make them with sticks and plasticine.

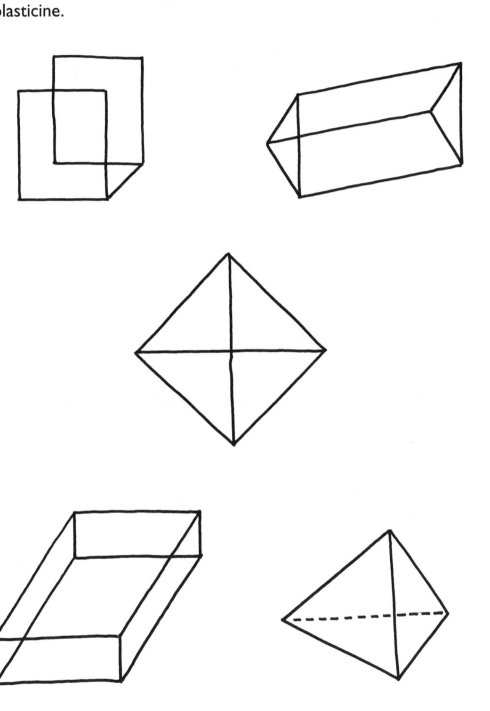

DOT-TO-DOT

Join the co-ordinate points on the grid to make a dot-to-dot picture.

(1, 8), (2, 9), (4, 7), (5, 7), (7, 6), (11, 2), (14, 3), (12, 1), (8, 2), (6, 3), (8, 1),
(7, 0), (8, 0), (9, 1), (8, 2), (6, 3), (6, 1), (5, 0), (4, 0), (5, 1), (4, 0), (5, 1),
(5, 3), (4, 4), (3, 3), (2, 3), (3, 4), (1, 4), (3, 5), (3, 6), (2, 6), (1, 7), (1, 8)

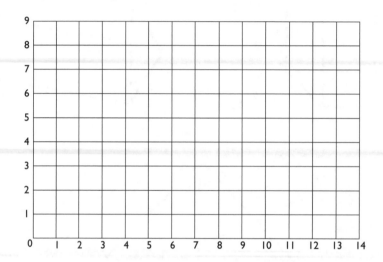

Make your own dot-to-dot. Draw a simple picture, using straight lines, on the grid below. List the co-ordinates you joined. Then give the list of co-ordinate points, in the correct order, to a friend to draw.

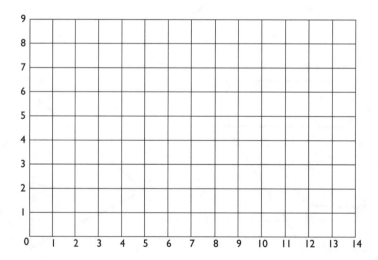

I joined the points:

A COUNTRY WALK

Look carefully at the graph, which shows the progress of a country walk.
Describe what you think happened.

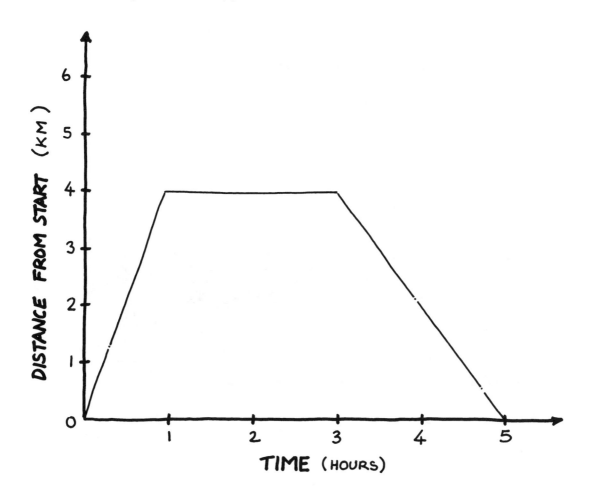

WHO'S WHO?

Which of these people fits which point on the graph?

Label each point with the initial of the person.

How old and how tall do you think each person might be?

Label the axes to show an appropriate scale.

SARA   MAX   BETTY   EMMA   SAM   KIM   DON

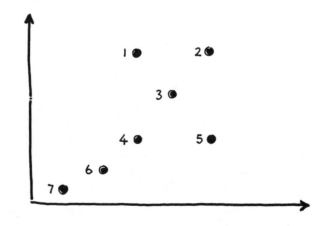

MY FAMILY

For homework, fill in the height and age of each member of your family on the graph below. Use a key to show who is who.

Start by marking yourself on the graph.

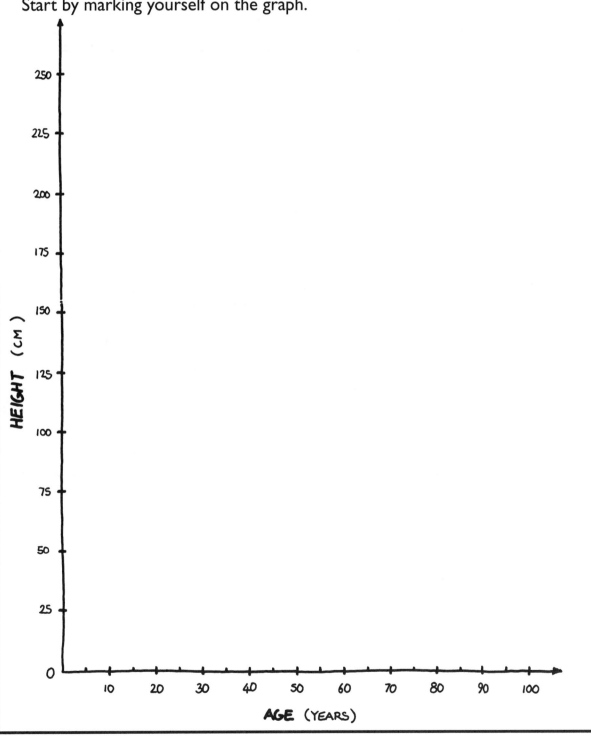

PLANNING FOR ALL LANGUAGE MODES

Topic:

ACTIVITIES	SPOKEN Listening/speaking	WRITTEN Reading/writing	ACTIVE Observing/doing	GRAPHIC Interpreting/ drawing, making

Comments:

REFERENCES
AND ANNOTATED
BIBLIOGRAPHY

This bibliography contains all references, together with details of a number of books which will be useful in implementing approaches discussed in this book. Children's books are listed separately after the general list.

Baker, John & Baker, Ann 1986. *From Puzzles to Projects: Solving Problems All the Way*, Melbourne, Nelson. (1993, Portsmouth NH, Heinemann.)
   This book provides a rationale for problem solving, practical advice on setting up a problem solving environment, and a range of problems.
Baker, Ann & Baker, Johnny 1991. *Maths in the Mind: A Process Approach to Mental Strategies*, Melbourne, Eleanor Curtain. (1991, Portsmouth NH, Heinemann.)
   The book describes twenty activities designed to stimulate mathematical thinking, and provides examples of the way children approach the problems.
Bell, A. 1986. 'Diagnostic Teaching 2: Developing Conflict-discussion Lessons', *Mathematics Teaching* 116.
   The characteristic of so-called 'diagnostic teaching' is to put students into situations which cause a conflict between their current beliefs and the new information. The conflict is (hopefully) resolved in small group discussion.
Burns, Marilyn 1982. *Math for Smarty Pants: or Who Says Mathematicians Have Little Pig Eyes*, Boston, Little, Brown and Company.
   The most capable readers and mathematicians in upper primary classes will enjoy reading and doing the activities in this book independently. Teachers can use the book as a wonderful resource for interesting and challenging mathematics games, problems and investigations.
Clyne, Margaret & Griffiths, Rachel 1990. *The Informazing Resource Book: Reading and Writing Nonfiction*, Melbourne, Nelson.

This book is a teachers' guide to the Informazing series of twenty-two books. Mathematics is integrated with science and language as you work through integrated units with the Informazing books

Clyne, Margaret & Griffiths, Rachel 1991. *Building on Big Books: Strategies for Using Texts*, Melbourne, Oxford University Press.

Ideas for using big books across the curriculum including mathematics.

Cooper, T., Atweh, B., Baturo, A. & Smith, R. 1993. 'Higher-intellectual functioning, empowerment and oral communication in mathematics instruction', in Stephens et al. 1993 (see below).

This chapter addresses two themes: how children can work effectively in small groups, and how teachers communicate their expectations and values to students.

Dalton, Joan 1985. *Adventures in Thinking: Creative Thinking and Co-operative Talk in Small Groups*, Melbourne, Nelson. (1992, Portsmouth NH, Heinemann.)

A very useful resource for all teachers wishing to get started on co-operative group work.

Del Campo, Gina, & Clements, Ken 1987. *A Manual for the Professional Development of Teachers of Beginning Mathematicians*. Melbourne, Catholic Education Office of Victoria and Association of Independent Schools of Victoria.

A manual developed to stimulate discussion of issues in mathematics teaching and learning in the early years of school.

Doig, Brian, & Masters, Geoffrey 1992. 'Through children's eyes: a constructivist approach to assessing mathematics learning', in Leder 1992 (see below).

This chapter presents some of the findings from the Basic Testing Program in New South Wales.

Donaldson, Margaret 1963. *A Study of Children's Thinking*, London, Tavistock Publications.

Donaldson's insights into children's thinking have definite implications for the ways we teach mathematics and other subjects.

Ellerton, Nerida & Clements, Ken 1991. *Mathematics in Language: A Review of Language Factors in Mathematics Learning*, Geelong, Deakin University Press.

A valuable resource for all teachers who wish to explore issues of language and mathematics. Australian and overseas research is reported and discussed.

Feynman, Richard 1986. *Surely You're Joking Mr Feynman*, London, Unwin Hyman. (1986, New York, Bantam.)

Feynman's autobiographical anecdotes are a great read.

Gooding, Anne & Stacey, Kaye 1993. 'How children help each other in groups'. In Stephens et al. 1993 (see below).

This chapter looks at the interactions of children working in small groups, and suggests some features of effective groups.

Griffiths, Rachel & Clyne, Margaret 1988. *Books You Can Count On: Linking Mathematics and Literature*, Melbourne, Nelson. (1991, Portsmouth NH, Heinemann.)

Besides providing integrated activities for forty books and poems, this book provides a rationale for the use of children's literature in learning mathematics, and an extensive annotated bibliography of children's books arranged thematically.

Griffiths, Rachel & Clyne, Margaret 1989, *Puffin Mathspack Middle and Upper Primary Level: Teachers' Notes*, Ringwood, Penguin Books.

A pack of twenty different children's books with a set of notes for teachers provide a wide range of integrated mathematics and language work.

Griffiths, Rachel & Clyne, Margaret 1990. *More Than Just Counting Books: Curriculum Challenges for Children*, Melbourne, Nelson.

Integrated activities, in mathematics, language, and other curriculum areas, for thirty-two counting books, and an extensive annotated bibliography.

Griffiths, Rachel, & Clyne, Margaret 1993. 'Real books and real mathematics'. in Stephens et al. 1993 (see below).

This chapter shows how using children's literature can be integrated with the development of mathematical skills and understanding.

Griffiths, Rachel & Clyne, Margaret 1994. *Maths Makes Sense: Teaching and Learning in Context*, Melbourne, Eleanor Curtain. (1994, *Math Makes Sense: Teaching and Learning in Context*, Portsmouth NH, Heinemann.)

This book explores contexts for teaching and learning mathematics. Many practical classroom activities, and a number of units of work, are provided.

Hughes, Martin 1986. *Children and Number: Difficulties in Learning Mathematics*, Oxford, Basil Blackwell. (1986, Cambridge MA, Blackwell.)

Following on from the work of Hughes and Donaldson documented in *Children's Minds*, Hughes explores children understandings and misconceptions about number. A fascinating and useful book.

Johnson, Terry & Louis, Daphne 1985. *Literacy through Literature*, Sydney, Methuen. (1987, Portsmouth NH, Heinemann.)

An inspiring and practical approach to integrating the learning of reading and writing with the enjoyment and understanding of literature.

Kerslake, D. 1979. 'Talking about mathematics', in R. Harvey et al., *Mathematics*, London, Ward Lock Educational. (quoted in Pimm, 1987, see below).

Leder, Gilah (ed.) 1992. *Assessment and Learning of Mathematics*, Hawthorn, Australian Council for Educational Research.

A valuable and interesting book which covers many aspects of assessment and learning, drawing on research from Australia, the US, the UK and The Netherlands.

Lovitt, Charles & Clarke, Doug 1988 *Mathematics Curriculum and Teaching Program Activity Bank* (MCTP), Melbourne, Curriculum Development Centre.

The two volumes of activities include lessons based on social issues, storyshells, sport, technology and many other contexts. The package includes video and computer support. A wonderful resource.

Lowe, Ian & Lovitt, Charles 1984. *RIME Pack A*, Melbourne, Department of Education, Victoria.

A series of mathematics lessons for year 7 students, which emphasise mathematical thinking, problem solving and relevance to students. Many of the ideas can be adapted for younger or older students.

Mottershead, Lorraine 1977. *Metamorphosis: A Source Book of Mathematical Discovery*, Sydney, Wiley.

A book of ten units on mathematics topics, including historical, recreational and creative aspects. While designed for secondary schools, there is much that can be adapted for upper primary use.

Pimm, David 1987. *Speaking Mathematically: Communication in Mathematics Classrooms,* London, Routledge. (1989, New York, Routledge.)

This book discusses the teaching of mathematics from the perspective of mathematics as a language. The nature of the mathematical register — vocabulary, syntax, symbols and structure — is explored.

Richards, Leah 1990. 'Measuring things in words: Language for learning mathematics', *Language Arts,* 67, 1, pp. 14-25.

The paper describes the wide range of writing activities in the author's mathematics classroom.

Room, Adrian 1989. *The Guinness Book of Numbers*, Enfield, Guinness Publishing. (1991, New York, Facts on File Inc.)

A mine of information about numbers, and written simply enough for many children in the upper primary years to understand and enjoy. Links are made with language, literature, sport and many other aspects of culture.

Shell Centre for Mathematical Education 1985. *The Language of Functions and Graphs*, Nottingham, Shell Centre.

This practical teaching kit is designed for secondary classes, but also has much to offer teachers of upper primary classes.

Skemp, Richard 1976. 'Relational understanding and instrumental understanding', *Mathematics Teaching*, 77.

In this classic paper, Skemp distinguishes qualitatively between rote learning, which enables the learner to do things, and relational understanding, which means that the learner can explain and make connections.

Skinner, Penny 1990. *What's Your Problem?: Posing and Solving Mathematical Problems in Junior Classes*, Melbourne, Nelson. (1991, Portsmouth NH, Heinemann.)

This book looks not only at solving problems but also at posing them. It includes short problems and longer investigations. The focus is on year 2, but the strategies and activities are applicable to a much wider age range. Use the ideas and activities with whole classes for discussion and comparison of methods, or with small groups and individuals who may enjoy challenges.

Stephens, Max, Waywood, Andrew, Clarke, David & Izard, John (eds) 1993. *Communicating Mathematics: Perspectives from Classroom Practice and Current Research*, Hawthorn, Australian Council for Educational Research.

This book brings together a wide range of perspectives on the language and mathematics issues, from Australian and New Zealand mathematics educators. A very useful and interesting reference.

Sullivan, Peter & Clarke, David 1991. *Communication in the Classroom: The Importance of Good Questioning*, Geelong, Deakin University Press. (1991, New York, State Mutual Books.)

This book provides clear and practical advice for teachers wishing to improve their questioning in order to help children towards a broader and deeper understanding of mathematics.

Swan, Malcolm 1988. 'Making mathematics more meaningful and relevant', in John Pegg (ed.), *Mathematical Interfaces*, Proceedings of the Australian Association of Mathematics Teachers Conference, Newcastle.

The paper describes some ways of making mathematics meaningful and relevant to primary and secondary students.

Swan, Malcolm 1982. *The Meaning and Use of Decimals*, Nottingham, Shell Centre for Mathematical Education.

>One of the many valuable Shell Centre publications, combining research findings and practical classroom activities.

Whitin, David, Mills, Heidi & O'Keefe, Timothy 1991. *Living and Learning Mathematics: Stories and Strategies for Supporting Mathematical Literacy*, Portsmouth NH, Heinemann.

>This exciting book describes a grade 1 classroom in which children deemed 'not ready for grade 1' were at the end of the year perceived as 'a gifted and talented class'. Building on children's interests and experiences, writing and drawing, using children's literature, and encouraging communication about all aspects of their learning were the major strategies used in this classroom.

Children's Books

Books referred to in the text are included here. These include narrative, nonfiction and interactive texts. Bibliographies of children's books which can be used for developing mathematical thinking are included in Griffiths & Clyne (1988, 1990), and Lovitt & Clarke (1988).

Allen, Pamela, *Mr Archimedes' Bath*, Harper Collins.
Allen, Pamela, *A Lion in the Night*, Puffin.
Anno, Mitsumasa, *Anno's Magical ABC*, The Bodley Head.
Anno, Mitsumasa, *Anno's Mysterious Multiplying Jar*, The Bodley Head/Putnam.
Baker, Jeannie, *One Hungry Spider*, Ashton Scholastic.
Base, Graeme, *The Eleventh Hour*, Penguin/Abrams.
Bogart, Jo-Ellen, *10 for Dinner*, Ashton Scholastic/Scholastic Inc.
Burningham, John, *The Shopping Basket*, Harper Collins.
Clement, Rod, *Counting on Frank*, Harper Collins/Gareth Stevens Inc.
Drew, David, *Alone in the Desert*, Rigby Heinemann.
Drew, David, *The Cat on the Chimney*, Rigby Heinemann.
Drew, David, *Creature Features*, Nelson.
Drew, David, *What Did You Eat Today?*, Nelson.
Drew, David, *The Book of Animal Records*, Nelson.
Faulkner, Keith, & Lambert, Jonathan, *The Python and the Pepperpot*, Pyramid Books.
Furchgott, Terry & Dawson, Linda, *Phoebe and the Hot Water Bottles*, Picture Lions.
Gackenbach, Dick, *A Bagfull of Pups*, Puffin/Houghton Mifflin.
Gretz, Susanna, *Teddybears 1–10*, Harper Collins/MacMillan.
Hutchins, Pat, *The Doorbell Rang*, Puffin/Morrow.
Kirsch & Korn, *Number Games*, Puffin.
LeSieg, Theo, *Ten Apples up on Top*, Harper Collins.
Room, Adrian, *The Guinness Book of Numbers*, Guinness/Facts on File.
Smith, Barry, *Cumberland Road*, Pavilion Books/Houghton Mifflin.
Viorst, Judith, *Alexander Who Used To Be Rich Last Sunday*, Angus & Robertson/MacMillan.
Walter, Marion, *Magic Mirror*, Tarquin.
Wheatley, Nadia & Rawlins, Donna, *My Place*, Collins Dove.
Williams, Kit, *Masquerade*, Jonathon Cape.

INDEX

Other books by Heinemann

Math Makes Sense
Teaching and learning in context
Rachel Griffiths and Margaret Clyne

Maths Makes Sense shows ways of teaching mathematics in contexts which make sense to children and therefore help children make sense of mathematics. This is a very 'hands on' book for teachers who want to do mathematics rather than read about the subject. Teachers are given examples of the principles and strategies at work in the classroom. These examples are accompanied by activities to use on the spot, so that teachers can see for themselves how exciting and effective teaching mathematics in context can be.
Contents: Teaching and learning in context and out of context *Changing beliefs about mathematics *Case studies *Planning contexts *Assessment and evaluation *Units of work: Maths Poems & Rhymes (K–3); Construction—A House for a Mouse (K–3); The Mice who Lived in a Shoe (Years 3–5); Trees (Years 3–5); Traffic Education (Years 5–7); Graphs (Years 5–7)
ISBN 0 435 08362 7   illustrated   144 pp

Responsive Evaluation
Making valid judgments about student literacy
Edited by Brian Cambourne and Jan Turbill

Changes in teaching practise caused by new understandings of how children learn have made many traditional methods of evaluation obsolete. Demands on teachers and educators to demonstrate accountability have put pressure on these same educators to devise new methods of assessment which demonstrate accountability and are appropriate to current teaching methods. Procedures must be established which lead to optimum learning, reflect holistic thinking, enrich classroom teaching and are seen to be rigorous, scientific and valid
Jan Turbill and Brian Cambourne have worked with teachers, principals, academics, parents and students to establish assessment procedures which fit these guidelines. They have all contributed to *Responsive Evaluation* and report on how they put the theory in to practise.
ISBN 0 435 08829 7   illustrated   144 pp

Raps & Rhymes in Maths
Compiled by Ann and Johnny Baker

A collection of traditional and modern rhymes, riddles and stories with mathematical themes, *Raps & Rhymes in Maths* can be used to provide a welcome break from more formal activities or can form the introduction or conclusion of a maths lesson. The raps, rhymes and stories provide openings for mathematical investigations and, most importantly, provide a source of enjoyment.
ISBN 0 435 08325 2   illustrated   90 pp

Counting on a Small Planet
Activities for environmental mathematics
Ann and Johnny Baker

Counting on a Small Planet uses mathematics to explore and explain how our actions affect the world we live in and develops children's awareness of the responsibility we each have for looking after our environment. The suggested investigations provide ideas for further action, and the Maths Fact Files at the end of each topic relate the local to the global situation.
Contents: All That Rubbish *Be Quiet *Water Use and Abuse *Not a Drop to Spare *When the Wind Changed *Design a House.
ISBN 0 435 08327 9   illustrated   98 pp

Maths in the Mind

Ann and Johnny Baker

Maths in the Mind looks at the acquisition of basic number facts and focuses on the development of mental skills and strategies within the context of broader activities. Twenty fully developed activities provide the children with the practise needed to develop, access and recall number facts speedily.
ISBN 0 435 08316 3   illustrated   120 pp

Mathematics in Process

Ann and Johnny Baker

Mathematics in Process investigates the purposes and conditions of learning and doing mathematics.
Part One looks at the child's experience and how children get involved, how young mathematicians work, how children communicate and learn from reflection.
Part Two sets out classroom approaches: identifying purposes for using and conditions for learning mathematics.
Part Three presents ideas on devising a curriculum, how to set up the classroom and features a complete section of activities to try immediately with your class.
ISBN 0 435 08306 6   illustrated   176 pp

Maths in Context

Deidre Edwards

Maths in Context shows how to integrate mathematics with the wider curriculum areas by using a central theme. This approach increases children's motivation, caters for individual differences and increases confidence in mathematical ability. Mathematics is seen as part of 'real life'.
A large section of the book presents ideas for activities based on Dragons, Our Environment, The Zoo, Party Time, Traffic, Christmas, Show and Tell, and The Faraway Tree.
ISBN 0 435 08308 2   illustrated   152 pp

Take-Home Science
Independent activities for science and technology
Jenny Feely

Take-Home Science provides a collection of stand-alone investigations, activities and experiments which children can work on at home or at school, by themselves or with a friend or a parent, to develop skills and knowledge in the areas of science and technology. Simple materials and equipment easily located at home or at school are used so that the idea of science and technology as part of the every day world is reinforced.
Guidelines for implementing the activities as part of a 'take-home' program are given, but the activities work well as part of the classroom program. The student activity sheets are highly illustrated to make the information and directions accessible to students of wide age and ability ranges. As part of recording and assessment procedures, children are required to keep a log book to record their findings after completing each activity. The log book is also an important way to communicate with the home. The program is designed for middle to upper primary children.
ISBN 0 435 08365 1   illustrated   128 pp